the Biology of Doom

the Biology of
Doom

The History of America's Secret Germ Warfare Project

ED REGIS

Henry Holt and Company   New York

Henry Holt and Company, LLC
Publishers since 1866
115 West 18th Street
New York, New York 10011

Henry Holt® is a registered
trademark of Henry Holt and Company, LLC.

Library of Congress Cataloging-in-Publication Data

Regis, Edward, date.
The biology of doom: The history of America's secret germ warfare
project / Ed Regis.—1st ed.
p.   cm.
Includes bibliographical references and index.
ISBN 0-8050-5764-1
1. Biological warfare—United States.   I. Title.
UG447.8.R44   1999   99-15024
355'.38'0973—dc21   CIP

First Edition 1999

Designed by Paula Russell Szafranski

Printed in the United States of America

1   3   5   7   9   10   8   6   4   2

For
Kevin Kelly

the Biology of Doom

Prologue

DUGWAY PROVING GROUND, UTAH
BW GRID NO. 4
TUESDAY, JULY 12, 1955
SUNSET

he test subjects are arrayed in a single straight line that stretches for more than half a mile across the desert floor. The file resembles a battle line from the old days of trench warfare, but this is only a field trial.

In a normal field trial, packs of caged mice, rats, guinea pigs, rabbits, or sheep inside wooden crates would be waiting patiently for the cloud of airborne infectious agent to wash over them. Sometime afterward—hours, days, weeks—most of the animals would show symptoms of the disease, and many of them would have died from it. Those that survived would be killed nevertheless, and all the test subjects would be autopsied to determine the precise medical effects of the biological agent used in the experiment.

But tonight's event will not be of the usual sort. Tonight, for the first time in the short history of the U.S. Army germ warfare program, a field trial using live infectious agents will include human beings as test subjects.

The thirty humans have been set out on the biological warfare test grid exactly as if they were animals, and in fact next to each small group of three human beings there is a cage of seven rhesus monkeys and a second cage of guinea pigs. The test line is thus a tidy biological cross section, fair and democratic, with no trace of species chauvinism anywhere in evidence.

Across from the line, in the middle distance, the last orange sun rays illuminate the top of Granite Peak, a 7,068-foot prominence. The air has cooled from the day's high of 97°F, and by the time of the experiment, which will occur after dark, the subjects will have been chilled by the dry breeze coming down from the mountains.

There are no human habitations within a radius of eighteen miles. From the nearest vantage point, Simpson Springs, an abandoned Pony Express station on a high bluff some ten miles to the east, the string of subjects is invisible to the naked eye. But no one is watching.

The sky overhead is clear and cloudless, a deep silent blue.

Soon it will hold a fine mist that will be carried along on the moving air toward the long line of humans and animals. The cloud will filter through their ranks like a fog, the mist produced by five type-C generators, spray nozzles that the Army germ warfare researchers have adapted from the Navy's E-4 marine mine.

The E-4 mine was one of the Navy's cleverer inventions. It was designed to be fired from a submarine's torpedo tube, but instead of speeding toward an enemy ship it would remain motionless under water for a specified period of up to two hours. Then it would rise to the surface. At that point, the mine's C generators would poke out above the waves and disseminate into the prevailing wind some forty-five quarts of a biological agent, a particular strain of virus or bacteria that would waft toward the enemy ship, and, depending on the species of microbe used, would either kill or incapacitate the crew. The mine would then scuttle itself and sink to the bottom forevermore.

On the theory that a spray device that worked well on the open ocean ought not to freeze up in the cold air of the high desert, the Army's germ warfare researchers have chosen the C generator for use tonight. Accordingly, five of them are arranged in an arc that focuses in toward the test subjects stationed 3,200 feet away. Once the nozzles start spewing out their supplies of live pathogen, it will take approximately four minutes for the aerosol cloud to cross the distance.

Earlier in the year, the Army had staged three dry runs of tonight's test, using guinea pigs as experimental subjects. In those trials, the researchers had learned how to calibrate the generators and aim the pathogen cloud at the center of the line, where the animals were bunched closely together in a section called the "dense sampling array." All the guinea pigs used in those trials, whether they were in the dense sampling array or not, had long since been sacrificed and incinerated, and from the postmortem examinations it was clear that many of them had been successfully infected with tonight's disease of choice, Q fever.

The "Q" in "Q fever" stood for Queensland, Australia, where the microorganism first appeared in 1935 as an epidemic among slaughterhouse workers. The men presented the typical symptoms of a flulike illness—fever, chills, headache, and so on—but blood tests failed to turn up a typical flu pathogen. Two researchers, MacFarlane Burnet, an Australian, and H. R. Cox, an American, independently discovered the microbe responsible. It proved to be an unusual type of rickettsia, and the organism was named after its two discoverers: *Coxiella burnetii*.

In the years since its discovery, *C. burnetii* had become known as a biological curiosity, a pathogen that produced no symptoms whatsoever in the vast majority of animal species that carried it. Among those it did affect, including sheep, goats, cattle, and guinea pigs, it caused spontaneous abortions in pregnant females about to give birth, but otherwise generally had little effect.

The disease was far worse in humans, among whom it produced fever, chills, and shivering; headaches that were severe and throbbing; eye pain, chest pain, cough, sore throat, weight loss, nausea, and vomiting; and a range of neurological problems including visual and auditory hallucinations. The symptoms could last for as little as a week or for as long as two to three months. A human victim was unlikely to die from Q fever, but deaths were not unknown, and about four percent of those who contracted the disease died from it. One out of thirty human cases died, on average, and there were thirty humans in the test tonight.

Other than its potential lethal effects, the worst part of Q fever was its long incubation period, which in humans ranged from ten to forty days. This meant that it could take a month or more for the first symptoms to show up, a month of watching and waiting, of oversensitivity to every stray pain.

Earlier in the evening, the Dugway technicians had filled each of the C generators with five ounces of infective slurry, the liquid that would shortly be streaming out of the spray nozzles. The microbes had been cultivated in a one-story yellow brick laboratory building at Camp Detrick, the Army's biological warfare research center in Frederick, Maryland, about fifty miles west of Washington, D.C. Laboratory technicians working in Building 434, the Virus Pilot Plant, had raised 3.5 liters of the *C. burnetii* microbe in embryonated chicken eggs. Then they extracted the liquid and purified it, transferred it into vials, placed the vials on dry ice, and shipped them by air to Dugway.

Now, finally, about an hour after sunset, the generators have been loaded and pressurized, and the air sampling devices on platforms next to the men are ready to sniff the aerosol as it comes wafting past. The samplers are powered by small vacuum pumps, and when the motors start running it means the test has begun.

After some initial erratic winds, the breeze is blowing steadily and from the right direction, from the generator arc toward the test line. Colonel William Tigertt, the Army medic in charge of the operation, walks along the line of men telling them in a calm voice, "When you hear the motors, just breathe normally. Remember to just breathe normally." No holding your breath, no gulping the air, no deep breathing exercises. This has to be a realistic test, approximating wartime conditions, when the enemy is not expecting an infectious puff of *C. burnetii* to float along on the night air. So when the time comes, *breathe normally.*

Tigertt now disappears into the control van, located well upwind and out of range, back behind the rank of C generators. The control van lights are dimly visible to the men, but otherwise the terrain is black.

And then, finally, the men hear the sound of the vacuum pump motors, a thin low drone.

The infective slurry is forced from the nozzles under the pressure of 250 pounds per square inch of carbon dioxide, and it billows out in a fine mist, like paint from a spray can. The particles are precisely sized, not so large that they'll fall to the ground like water from a hose, not so small that the subjects will exhale them back out again like cigarette smoke. When the men breathe them in, some of the particles will remain behind in the lung.

The spray nozzles are five feet above the ground, the same height as the human nose. The individual lines of spray are caught by the moving air; they travel downwind and merge, now forming a coherent cloud. It drifts toward the center of the test line, toward the dense sampling array, the bunched groups of guinea pigs, monkeys, and men.

Here it comes! . . . Here it is! . . . You can feel it on your face!

A fog. A vapor. A soft wet mist.

part one

1

wenty years earlier, the U.S. Army would have laughed at the idea of making war upon the enemy by spraying germs out of a can. The Army's official position, in those days, was that biological warfare was science fiction.

That position had been formulated, argued, and loudly proclaimed by one Leon A. Fox, M.D., a major in the U.S. Army Medical Corps. Fox viewed himself as a great debunker of popular delusions, and in 1932 he had written an article for the Army journal *Military Surgeon* in which he took a dim view of the entire germ war fantasy.

"Bacterial warfare is one of the recent scare-heads that we are being served by the pseudo-scientists who contribute to the flaming pages of the Sunday annexes syndicated over the nation's press," he wrote. "I consider that it is highly questionable if biologic agents are suited for warfare. Certainly at the present time practically insurmountable difficulties prevent the use of biologic agents as effective weapons."

For one thing, most of the world's microbes were highly perishable: they were destroyed by heat, cold, and even by sunlight. Placing germs inside weapons and firing them off like cannonballs pretty much guaranteed that the organisms would be dead on arrival.

"Shells can be used to project missiles and chemicals on to an enemy many miles distant," he said, "but bacteria cannot be used in this way. No living organism will withstand the temperature generated by an exploding military shell."

You could spread bacteria from an airplane, he conceded, just as a crop duster sprayed insecticide over farmland. Still, "their effect would be quite local and probably less dangerous and less certain than high explosives used in the same way."

And in any case there was the boomerang effect to consider. Unless you've immunized your own army against the specific biological agent you're going to use on the enemy, that agent could come back and kill you as easily as them. If, on the other hand, you've immunized your own forces against certain bacteria, then the other side could do so as well, making the entire biological assault an exercise in futility.

But if you wanted to wage war with germs even in spite of all these obstacles, exactly which diseases would you try to inflict upon the adversary?

Not meningitis, for the organism was "so delicate that even on the most favorable culture media it rapidly dies when exposed for even a few hours to temperatures much below that of blood heat." Smallpox was out because soldiers were immunized against it as a matter of course. Influenza was a possibility, except for the fact that no one knew how to start a proper epidemic of the disease: the microbe was always skulking around in the environment somewhere, and it broke out sporadically and at random for unknown reasons. Handling the virus in a controlled manner and directing it upon the foe at will seemed entirely out of the question.

Plague was excluded by the boomerang argument. "The use of bubonic plague today against a field force, when the forces are actually in contact, is unthinkable for the simple reason that the epidemic could not be controlled. The torch once set off might destroy friend and foe alike, and would therefore prove of no value as a military weapon."

Botulinum toxin sounded like a good weapon, and indeed plenty of Sunday supplement writers seemed to get hysterical over it, claiming that an ounce or so of the substance would be enough to kill every man, woman, and child on earth. Mathematically that might be true, said Fox, but it was of no account if you couldn't physically parcel out a minute portion of that ounce to each of those persons one by one. A bare mathematical possibility, in other words, was not the same thing as a genuine material prospect. "There were over one hundred billion bullets manufactured during the

World War—enough to kill the entire world fifty times," Fox said. "But a few of us are still alive."

But for all the cold water he threw on the idea of biological warfare in general, even Leon Fox had to admit that there was one biological agent out there that approximated to a high degree what could be called the "perfect military pathogen." This was anthrax.

Anthrax was a spore-forming microbe, meaning that when the bacillus was thrust into unfavorable conditions it curled itself up in a tiny ball and built around its outer surface a capsule that amounted to a hard hide. Such spores were known to be remarkably stable and resistant to the destructive influences of light and heat, and they could remain that way, with no loss of virulence, for a period of many years.

"These spore-forming invaders are a real problem," said Leon Fox. "We cannot dismiss anthrax so readily."

He also made one other concession, concerning bubonic plague. Even though, because of the boomerang problem, it was of no use in close quarters on the battlefield, plague could still be used, he said, "to harass civil populations." This would be true especially if the attacker could leave the area right after introducing the microbe.

He even had an idea about how to do this: "It may be possible for airplanes flying low to drop recently infected rats."

Leon Fox's piece was published in the March 1933 issue of *Military Surgeon* under the title "Bacterial Warfare: The Use of Biologic Agents in Warfare." Shortly after it appeared in English, a Japanese translation was prepared in Tokyo, where it was read by one Shiro Ishii, a physician in the Imperial Japanese Army.

Ishii was about the same age as Fox, and was also a major in his country's Army Medical Corps. However, Ishii was not as skeptical about the idea of germ warfare as Fox was. In fact, he regarded Fox's piece as "fantastic" and "not based on scientific facts."

Ishii's own view was that germ warfare was a distinct possibility. Why had it been outlawed by the 1925 Geneva disarmament convention, he reasoned, unless it posed a realistic threat to modern armies? Organized states did not go to all the trouble of banning forms of warfare that had little or no chance of working, but on June 17, 1925, in Geneva, representatives of twenty-nine nations (including the United States) had signed a "Protocol for the Prohibition of the Use in War of Asphyxiating, Poisonous or other Gases, and of Bacteriological Methods of Warfare."

Gas warfare, Ishii knew, had caused a million casualties in World War I. Although bacteria had never been used as weapons, it was obvious they could do a lot of damage. Every military physician understood that during all the wars of past history, more men had been killed by disease than by actual battle. Malaria, dysentery, cholera, typhus, bubonic plague, and other diseases had devastated countless armies. Why couldn't a field commander capitalize on that fact, turning nature's own prefabricated agents of destruction into controlled and directed offensive mechanisms?

Ishii had gotten some firsthand experience of the murderous potential of epidemic diseases when in 1924 he waded into an outbreak of an unknown pathogen on the Japanese island of Shikoku. Patients were losing weight, shaking with chills, and many became partially or totally paralyzed by the infection. Before long, 3,500 people had died of severe brain inflammation.

Much later, the cause was found to be Japanese B encephalitis virus, a microbe that was transmitted to humans by mosquitoes. The outbreak had natural causes and had not been artificially induced. Still, the epidemic was exactly what a germ warfare attack would look like: a strange disease appearing out of nowhere all at once, swamping the health care system and leading to many casualties. It was a formative experience for Shiro Ishii.

As a person, Ishii was overbearing and generally obnoxious. At Kyoto Imperial University, where he'd gotten his medical degree in 1920, he made a habit of coming into the lab late at night, running through a succession of test tubes, beakers, and other lab glassware, and then leaving the mess for others to clean up. Nevertheless, he ingratiated himself with his superiors, married the university president's daughter, and fathered a large brood. He spent his after-hours in bars and geisha houses, having a whale of a time. He had a mesmerizing effect on people, even strangers, and was said to have a hypnotic appearance.

In 1928, Shiro Ishii left Japan and toured the world, heading south to Singapore, then west to Ceylon, Egypt, Greece, Turkey, then moving systematically through Europe, hardly missing a single country. He crossed the Atlantic, visited the United States and Canada, and returned to Japan via Hawaii. A raft of legends soon sprang up about him. One dating from this period claims that he lived in Boston for a time and studied germ warfare at MIT, but MIT had no such program.

Two years later Ishii was back in Japan and working as professor of immunology at the Tokyo Army Medical College. While there he made his one positive contribution to military medicine, inventing a ceramic filter for water purification. The filter removed all kinds of impurities—bacterial, viral, or chemical—without boiling or any sort of chemical treatment.

Such a device, if it worked, would be a boon to modern armies, saving them from recurrent epidemics of waterborne diseases. Supposedly, Ishii performed live demonstrations of his device by urinating into the filter and drinking the output, and there was a legend that he repeated this display before Emperor Hirohito and invited him to sip the discharge. The filter was apparently a success because both the Japanese Army and Navy adopted it for field use, and a Tokyo firm manufactured it in various sizes, paying Ishii substantial royalties in the process.

By 1931, then, Ishii was a miracle man who strode into breaking epidemics, had circled the globe, invented an appliance that purged water of evil influences, and got rich on the proceeds. So when he gave impromptu lectures about the advantages of using germs as weapons, people paid attention. Indeed, his reasoning appeared to be unanswerable: microbes made people sick and killed them, and they did so reliably and according to the known laws of microbiology. Germs were invisible, cheap, and easy to grow in quantity. Why not utilize their offensive potential? And why not make him, Shiro Ishii, leader of the whole project?

In 1932 the Japanese Army gave Ishii a research laboratory at the Army Medical College in Tokyo, a bacterial production facility in Harbin, China, and a test site in the nearby rural village of Beiyinhe. Three separate institutions devoted to biological warfare research and all of them under the control of the same man, Shiro Ishii.

And all of it while Leon Fox insisted that germ warfare was a pipe dream.

In August 1933, a year after Shiro Ishii became the emperor of Japanese germ warfare, and the same year in which Leon Fox published his piece dismissing the notion, the Germans were staging a series of practical, hands-on biological warfare experiments in the ventilation shafts of the Paris Metro and in the tunnels of the London Underground—if an article that appeared in a staid British periodical *The Nineteenth Century and*

After could be believed. The piece, "Aerial Warfare: Secret German Plans," by the British journalist Wickham Steed, published in the July 1934 issue, told an amazing story.

Recently, by means he did not specify, Steed had received a cache of secret German documents. One was a memorandum allegedly written in Berlin in July 1932, by an unidentified official in the gas warfare division of the German War Office. The document described how bacteria then known as *Micrococcus prodigiosus* (later renamed *Serratia marcescens*) were commonly used in medical schools to demonstrate the airborne transmission of infectious diseases. The *Micrococcus prodigiosus* bacteria, thought to be harmless to humans, had a bright red hue—on culture plates they appeared as tiny red specks—and so to provide an object lesson in how bacteria could float through the air, a medical school lecturer would place a small quantity of the stuff in his mouth, mix it with his own saliva, and then proceed with the day's talk. At the end, the speaker would collect the culture plates that he'd earlier placed at various points around the room and incubate them overnight. Next day, lo and behold, fresh new colonies of the red bacteria had grown up on the culture plates. The conclusion was obvious for all to see: if the lecturer had been ill with tuberculosis or some other communicable disease, the causative agents would be flying around the room and infecting everyone within range.

The other lesson was that *Micrococcus prodigiosus* microbes made excellent biological tracers, and could be used to track air currents in places other than in medical school lecture halls. That gave the German gas warfare official an idea: Why not use that same bacillus to trace the airflow patterns into and throughout the subway tunnels of London and Paris? Then you'd know the probable results of spraying the subway air vents with chemical gases or pathogenic bacteria. The memo concluded, "If these bacilli could be successfully rained down from an aeroplane, with sufficient concentration, from various heights and in varying conditions of wind and weather, etc., and, as in the case of the medical demonstrative experiments just mentioned, could be caught by culture plates on the ground, then one could study at one stroke, aerodynamically and meteorologically, not only bacteriological but also chemical spraying."

The subways could be prime targets in a future war, especially if Londoners and Parisians flocked to the tunnels during air raids. With the intake air contaminated by anthrax or other bacteria, the underground refuge would be converted into an incubator of a mass epidemic.

In 1933, according to Wickham Steed's documents, German agents had already sprayed billions of the bright red *Micrococcus prodigiosus* microbes into the Paris Metro system by driving a car several times around the Place de la Concorde subway entrances while releasing the bacteria. The car's exhaust gases had disguised the aerosol cloud, and no one had noticed the mock attack.

Six hours later, at the Place de la République Metro station, a mile and a half from the dispersal point, the German covert agents found that more than four thousand colonies of the *Micrococcus prodigiosus* germs had been deposited on the bacterial culture plates they had set out in the station. For biological warfare purposes, that was a favorable result: it meant that you could fly over the Place de la Concorde, which was easy to spot from the air, day or night, drop a biological bomb, and be confident that the bacteria would seep down into the subway system and infect anyone there or in the immediate neighborhood. The Germans, according to Wickham Steed, had run similar spray tests in London, using the Piccadilly Circus Underground station as the attack point. The results had been similar to those in Paris.

No one ever proved, then or afterward, whether the "secret German documents" were real or phony, fact or fantasy. But it made no difference, the underlying message was equally disturbing either way: subways were vulnerable to biological organisms dropped from the skies. Worst of all, there seemed to be no obvious or easy way to defend against it.

Steed's article created a mass sensation in the United Kingdom. Politicians made speeches about it in Parliament, public health officials considered possible responses to a biological assault on the Underground, and military leaders examined various counterattack strategies, it being an article of faith among them that the best response to a given threat was always retaliation in kind. The cities of Berlin and Hamburg had their own subway systems, and they, too, were open to bacterial attacks. There seemed to be only one logical recourse: if the Germans were preparing for germ warfare, the British would not be far behind.

In 1939, some five years after the Wickham Steed affair, a Japanese physician by the name of Ryoichi Naito showed up, unannounced, at the Rockefeller Institute for Medical Research in New York.

The Rockefeller Institute was located on a rich and fabulous fourteen-acre campus between 64th and 68th Streets on the east side of the city. The institute's walks were bordered by tall trees, fountains, ponds, and patios, making the school reminiscent of an Italian villa in Tuscany. Into this heavenly setting, on February 23, 1939, came Dr. Naito, who was an assistant professor at the Army Medical College in Tokyo, the very place where, as it happened, Shiro Ishii ran a bacterial warfare research lab. Naito carried with him a letter of introduction from the military attaché of the Japanese Embassy in Washington explaining that medical researchers in Japan needed samples of yellow fever virus for the purpose of creating a vaccine.

Yellow fever was a viral illness characterized by high fever and headache, jaundice, black vomit, and bleeding. It was transmitted by mosquitoes and common in the tropics. There was essentially no treatment for it, as was generally true of diseases caused by viruses, and the mortality rate was approximately five percent.

Naito met with Dr. Wilbur A. Sawyer, director of the Rockefeller Institute Virus Laboratories, and told him that he'd come to the institute precisely because one of its scientists, Max Theiler, had recently developed a yellow fever vaccine. The Japanese needed samples of the virus, said Naito, in order to make a vaccine of their own.

Sawyer listened to this with some doubts. For one thing, yellow fever was a disease of the tropics, mainly Africa and South America, whereas there was little if any of it to be found in Japan. Why, then, would they need to protect themselves against it?

But there was a second consideration. To prevent the emergence of the disease in areas where it didn't exist already, both the League of Nations and a Congress of Tropical Medicine had passed resolutions prohibiting the importation of yellow fever virus into Asian countries for any reason. Sawyer therefore was regretfully unable to provide any samples of the virus to Dr. Naito.

Three days later, on a Sunday morning, a Rockefeller Institute technician who worked in Sawyer's virus lab, a man by the name of Glasounoff, was stopped on the street by an unidentified man who spoke with a foreign accent. The stranger was around forty, had a small mustache, and was dressed in a blue pinstripe suit and brown overcoat and hat. He told Glasounoff that he needed samples of the *Asibi* strain of the yellow fever virus—an extremely virulent strain, lethal for humans—for a scientific project in Japan. He couldn't ask the lab chief directly, he said, because there was some professional rivalry involved, a case of the unfortunate competition that so often prevented progress in the sciences. But if Glasounoff could provide the samples, he would be paid a thousand dollars.

This was highly unusual, and Glasounoff declined the proposition, whereupon, even more alarmingly, the foreigner upped the offer to $3,000—$1,000 now, plus $2,000 on delivery. Glasounoff refused, got away, and reported the event to Dr. J. H. Bauer, his boss at the virus lab.

Bauer sent news of the incident up through the chain of command at the Rockefeller Institute, and a full report of the occurrence eventually reached the State Department in Washington. But nobody there knew what it meant.

On October 4, 1940, some twenty months after the two failed Japanese attempts to get yellow fever virus samples from the Rockefeller Institute, and eight years after Leon Fox had proposed the idea of dropping "recently infected rats" as a way of causing a plague epidemic, a lone Japanese plane circled the town of Chuhsien in Chekiang Province south of Shanghai, China.

The plane made a low pass over the western district and dropped out an indistinct cloud of stuff, as if the pilot had released a bucket of sand. On the ground, a man by the name of Hsu saw the material filter down through the air and settle on the street in front of his house. He stooped down to examine the particles and saw that they consisted of wheat grains and rice; he also noticed a large number of fleas crawling around in the mixture. He swept up a sample of the grains and fleas and brought it to the local air-raid station, whose personnel forwarded it to the provincial public health laboratory.

The public health laboratory workers knew that rat fleas were the normal carriers of the bubonic plague bacillus, among other things, and so they crushed the fleas and tried to culture out any bacteria they might be harboring. All their attempts to culture an organism from the fleas failed, but on November 12, thirty-eight days after the Japanese plane dropped its cargo over the city, an epidemic of bubonic plague broke out in Chuhsien in the area where the grains and fleas had been found in abundance.

The epidemic lasted for twenty-four days and resulted in twenty-one deaths. According to local health records, the town of Chuhsien had never before experienced so much as a single case of bubonic plague.

The Japanese plane had apparently caused the epidemic. The wheat and rice grains it dropped had attracted the local rats, the rats attracted the fleas, and the rats carrying the fleas brought the bacillus into village homes in the classic manner of plague transmission.

The case for all this would have been clinched had the lab workers performed an animal inoculation test, crushing the fleas, injecting some of the contents into lab rats or guinea pigs, and causing them to become ill with the disease. The provincial public health laboratory, however, lacked the facilities to perform such an experiment.

But the identical scenario was soon repeated elsewhere, in the port city of Ningpo, also in Chekiang Province, where on October 27, 1940, Japanese planes staged another bombing raid, scattering a large quantity of grain in the process. Nobody took any samples, but bubonic plague broke out in Ningpo just two days later. The epidemic lasted thirty-four days and claimed a hundred human lives.

Then on November 28, 1940, with the other two epidemics still in progress, three Japanese planes flew over Kinhwa, a city midway between Chuhsien and Ningpo. Afterward, the ground was covered with tiny pearly-white granules about the size of shrimp eggs. The residents collected the pellets and brought them to the local hospital.

The objects were roughly a millimeter in diameter, about the size of the roller on a ballpoint pen. They were translucent and when placed in water they swelled up and broke open, releasing a substance that formed a milky suspension.

A technician smeared the suspension onto a glass slide and placed it on the focusing stage of a light microscope. The objects were somewhat fuzzy and blurry but recognizable nonetheless. The plague organism had a distinctive bipolar appearance: it looked like a safety pin, with two dark ends separated by a thin clear shaft. That was the shape that now appeared in the microscope.

On the other hand, the laboratory failed to culture the bacillus from the samples, and no bubonic plague epidemic ever broke out in Kinhwa. And so after three consecutive attempts there was only circumstantial evidence that the Japanese were performing biological warfare experiments in China.

The Japanese provided more evidence a year later when at 5 A.M. on November 4, 1941, a solitary plane flew over Changteh, a city in Hunan Province some 500 miles inland from the other three sites. Air-raid sirens went off and people took cover as the aircraft made three low passes down Kwan-miao Street. Wheat and rice, small flecks of paper, several pieces of

cotton wadding, and miscellaneous other objects fell from the sky like con-
fetti. The locals collected these strange "gifts from the enemy" and took
them to the police, who brought them to the Presbyterian hospital. There
lab workers looked through the microscope and saw what they thought
was the plague organism, although they performed no corroborating tests.

But a week after the air raid, an eleven-year-old girl who lived on Kwan-
miao Street developed a high fever and swelling of the lymph nodes, the
"buboes" from which bubonic plague got its name, and she was dead within
two days of the first symptoms. There had never been any cases of bubonic
plague in Changteh for as far back as anyone could remember.

A year after the first Japanese plague attacks on China, the Americans
finally decided to hold discussions about the potential threat of biological
warfare.

Word of the yellow fever incident at the Rockefeller Institute had
reached Lieutenant Colonel James S. Simmons of the U.S. Army Surgeon
General's office. Simmons was one of the few in the Army who had rejected
Leon Fox's skepticism about the possibilities of germ warfare. In fact,
Simmons had argued as far back as 1937 that the Japanese could easily create epidemics in the United States by dropping swarms of yellow-fever
infected *Aedes aegyptii* mosquitoes on American shores. So when he heard
in January 1941 that the Japanese had been scouting around for samples of
the virulent *Asibi* strain and were willing to pay good money for it, he
thought he knew exactly what was happening, namely, that they were trying to make a biological weapon out of the yellow fever virus.

Still, nothing happened for several months. In August 1941, finally,
Simmons wrote to the secretary of war, Henry L. Stimson, telling him that
certain segments of the U.S. Army were now prepared to give more cre-
dence than formerly to the whole notion of biological warfare. The orthodox view of germ warfare, which was that it was an unworkable phantasm,
was based largely on theory, not on hard evidence or experiment. Simmons
suggested to Stimson that the War Department ought to take the matter in
hand, by undertaking a research program to create possible defenses as
well as to prepare to reply in kind to a biological attack.

In time-honored government manner, the secretary of war responded by
creating a committee. Composed of nine of America's top biologists, from
Johns Hopkins, Yale, Cornell, the Rockefeller Institute, and the University

of Chicago, and headed up by Edwin B. Fred, professor of bacteriology at the University of Wisconsin, the WBC Committee itself was never sure what the letters "WBC" stood for. Officially and for the record they meant "War Bureau of Consultants"; unofficially they were a deliberate backwards spelling of "Committee on Biological Warfare." Whichever it was, the members met for the first time on November 18, 1941, at the National Academy of Sciences in Washington. They decided they would perform a literature search.

Nine days later, the U.S. Army received its first reports of the Japanese bubonic plague attacks on the Chinese village of Changteh.

Ten days after that, on December 7, 1941, the Japanese bombed Pearl Harbor, at which point the committee members decided to take the prospect of germ warfare somewhat more seriously than before.

Two months later, on February 17, 1942, the WBC Committee issued its first formal report. By any standard it was an impressive document, more than two hundred pages long, with thirteen appendixes including an annotated bibliography that ran to eighty-nine pages.

The results of their literature search were especially surprising. Beyond the narrow confines of the U.S. Army, the world of science was full of proposals for the intentional dissemination of noxious microbes as a means of killing or incapacitating the enemy. Sixteen articles discussed the use of animals as means of spreading bacteria, while fourteen proposed using insects. Sixteen additional articles mentioned dissemination by airplane, fifteen talked about sabotaging public water supplies, ten suggested bacterial aerosols, and six proposed bacterial bombs.

The old arguments against germ warfare were now seen to be invalid in every respect. "This type of warfare has been frequently rejected as impracticable, or unlikely to be used because of the 'double-edged' nature of the weapon," the committee members said. But the "boomerang" effect was a problem only for close-range fighting; newer methods of long-range attack—dissemination by airplane, for example—entirely removed that obstacle.

Leon Fox's contention that bacteria couldn't survive being dropped in bombs was now revealed to be mere conjecture. Maybe they could survive, but who really knew without making the attempt? This was an empirical matter that could be settled only by experiment.

The committee members also had several practical suggestions as to which specific germs and diseases might be applied to the task. "Meningococcal meningitis might be spread by spraying meningococci in crowded quarters," they said. "Typhoid could be introduced by sabotage into water and milk supplies and by direct enemy action into reservoirs. . . . Botulinus toxin might be conveyed in lethal amounts through water supplies. . . . Plague could be introduced into any of the large cities or ports by releasing infected fleas or rats. . . . Diphtheria can be spread by dissemination of cultures in shelters, subways, street cars, motion picture theaters, factories, stores, etc., by surreptitiously smearing cultures on strap handles and other articles frequently touched."

They spoke of relapsing fever and hemorrhagic jaundice; of smallpox, rabies, and poliomyelitis; of Rift Valley fever, dengue, and influenza. All of them were potential weapons—and that was only for human beings. Animals and plants had diseases of their own: Newcastle disease, fowl plague, foot-and-mouth disease, hog cholera, rice blast, cereal stem rust, wheat scab, late blight of potato—all these things were out there in nature, ready and waiting for conversion into offensive weapons.

The committee's most imaginative suggestion, though, was to combine many diseases into a single live delivery package, a living buzz bomb of assorted pathogens that could be sent winging its way toward man or animal. A mosquito, they thought, would be perfect for the task, and so they recommended that "studies be made to determine whether mosquitoes can be infected with several diseases simultaneously with a view to using these insects as an offensive weapon."

Clearly, the committee members had warmed to their subject and were now firm believers in this new and exotic form of waging war. "Biological warfare is regarded as distinctly feasible," they said. "We are of the opinion that steps should be taken to formulate offensive and defensive measures."

A few months later, in June, the WBC Committee issued its second and final report. "The best defense for the United States is to be fully prepared to start a wholesale offensive whenever it becomes necessary to retaliate. In biological warfare," they added with emphasis, *"the best defense is offense and the threat of offense."*

The last line of the report, and the final conclusion of the War Bureau of Consultants, was: "Unless the United States is going to ignore this potential weapon, steps should be taken immediately to begin work on the problems of biological warfare."

Its job done, the group disbanded.

2

he irony was that while the U.S. Army, and even the nine WBC Committee members, imagined that they were on top of everything pertaining to science and technology and on cutting edge of military research, the fact was that insofar as biological warfare was concerned they were several years behind the times. Not only were they ten years behind Japan, they even trailed their own allies. It was bad enough that the British were far ahead of the Americans, but at least it was understandable: they were in Europe, after all, where the fighting was. Even Canada, however, was ahead of the United States in germ warfare research and development.

The Canadians had a germ warfare project under way by 1940, and it hadn't even been started by the government. It had been motivated by a country doctor and funded by a group of private investors.

The doctor was Frederick Banting, a World War I veteran who had a small medical practice on the outskirts of Toronto. In the 1920s he became interested in diabetes, a condition in which the body was unable to break down ordinary foodstuffs and convert them into energy. He got some lab space at the University of Toronto, did experiments on dogs, and soon discovered the hormone that controlled carbohydrate metabolism: insulin. For this discovery he and John Macleod, who ran the laboratory, shared the

1923 Nobel Prize in medicine or physiology, the first Nobel ever awarded to a Canadian.

Banting was of a highly independent turn of mind, and in 1939, at the beginning of World War II, he wrote in his diary that the coming conflict would be "a war of scientist against scientist. This war above all in history will be one in which the application of science to warfare will give one side or the other the advantage."

One application of science, he decided, was the use of bacteria and viruses as offensive weapons. In past wars, germs had always killed more soldiers than fighting, and it was only logical to capitalize on the agents of destruction that were already out there for free. He himself was full of ideas about how to utilize them: You could rear swarms of disease-carrying insects and disperse them over cities. You could infect bullets, bombs, and artillery shells with bacteria before firing them at the enemy. You could mix viruses with sawdust and disperse the particles from high-flying aircraft.

Neither Canada nor Mother England, however, had the wisdom to grasp these facts. "The most serious situation that faces our dear, old England," he wrote in his diary, "is that of bacterial warfare. The sad part is that we have not the means to retaliate. England suffers from a plethora of senile brains that live in the late Victorian period." As for Canada, "our government is led by a senile fossil of a vintage that would do credit to whisky."

But then, in the summer of 1940, Banting met an official in the Canadian government, James Duncan, who had contacts in the business world. Duncan had been vice president of Massey Harris, the Canadian tractor firm, and he was friendly with any number of rich capitalists.

He, for one, was wholly persuaded by Banting's arguments, and out of the blue he suggested that maybe one or two of his business friends might be willing to advance some funds—in the form of donations, contributions, patriotic gifts—to be used as seed money for work on biological warfare, toxic gases, and whatever other projects Banting might have in mind.

Soon John David Eaton, head of the T. Eaton Co. department store chain, had pledged $250,000 to Banting's cause. Then Sir Edward Beatty, president of the Canadian Pacific Railway, promised to contribute the same amount. And Sam Bronfman, head of the Seagrams liquor empire, matched the offers of the other two men, dollar for dollar. But there was much more than that to come from other Canadian businessmen, and by the end of this grand outpouring of corporate cash, they had donated over $1.3 million to Canada's developing germ warfare slush fund.

They gave the money to the National Research Council, the Canadian government's scientific research, development, and funding organization, which formed a group to control the assets and placed two of the private benefactors, John Eaton and Sam Bronfman, on the board as members. In July 1940, well before the U.S. Army started paying any serious attention to germ warfare, the group funded a bacterial warfare research program at the Connaught Laboratories of the University of Toronto, thereby making Canada, not the United States, the birthplace of formal germ warfare work in North America.

In their Toronto labs, Banting and others now considered the use of the psittacosis microbe, the cause of "parrot fever" in birds, as well as fever and respiratory ailments in humans, as a biological weapon. On or about October 8, 1940, Banting and his colleagues went up to Balsam Lake northeast of Toronto to test the idea of using sawdust particles as a disease-carrying medium. As a plane flew a zigzag pattern overhead and spewed out several different grades of uninfected sawdust, Banting on a boat and others on the shore used surveying instruments to measure the size and shape of the resulting particle cloud.

Neither Banting nor any of his colleagues had ever heard of Shiro Ishii or knew of his plans for dropping rice, grains, and infected fleas upon the enemy. Yet Banting's experiment over Balsam Lake occurred just four days after Japanese aircraft dropped plague-ridden fleas over the town of Chuhsien in the Chekiang Province of China.

Two years later, in November 1942, almost a year after Pearl Harbor, the Americans had gone so far as to form yet another committee for the purpose of investigating biological warfare. The new group, formally known as the War Research Service, was the official successor agency to the WBC Committee, whose report the previous February was supposed to have started the ball rolling. All it had actually done was to spawn a succession of new committees, not only the War Research Service itself, but also a separate ABC Committee, and later even a DEF Committee—so many committees they'd apparently run out of names.

Chairman of the War Research Service was George W. Merck, a civilian chemist and president of Merck & Co., the pharmaceutical manufacturing firm with headquarters in Rahway, New Jersey. Merck was forty-eight years old, blond, and blue-eyed, and at six feet five inches tall was a veritable slab of marble. He'd been appointed the czar of the American biological warfare program in August and had immediately moved into his new offices at the National Academy of Sciences in Washington.

The American president, Franklin D. Roosevelt, had approved a biological warfare research program a few months before, in May, after his secretary of war, Henry L. Stimson, had received the first formal report of the WBC Committee. On April 29, Stimson had written to Roosevelt, saying: "Biological warfare is, of course, 'dirty business,' but in the light of the committee's report, I think we must be prepared." As much as Roosevelt himself detested chemical warfare (and had pledged not to use it unless it was first used by the enemy), he nevertheless saw the wisdom of being prepared to retaliate in kind if so attacked, an argument that would apply to biological weapons as well, if and when they were ever developed.

At this point, however, biological weapons were still in the future. The American scientists, for all their various committees and study groups, had yet to grow a single bacterium for offensive purposes, perform a dissemination experiment, or infect any animal with a pathogen spray.

Now, in November, two British scientists, Paul Fildes and David Henderson, came to Washington to see Merck and a few top officials of the U.S. Army Chemical Warfare Service, the agency responsible for biological weapons research. What Fildes and Henderson told Merck and the others would finally launch the United States onto its quarter-century-long biological warfare program.

Paul Fildes (who pronounced his name to rhyme with "wilds") was England's top bacteriologist and the head of the biology department at Porton Down, England's chemical and germ warfare research institute. Fildes was fifty-eight, bald, blunt, and taciturn, and walked along so deliberately it was as if he had flippers on his feet. He was a lifelong bachelor with a hatred of bureaucracy, and an implacable enemy of all forms of humbug and pretension. He had gone to the London Hospital Medical College in 1904 with the hope of becoming a surgeon, but he soon discovered that the ways of microbes were of far more interest to him than the mechanics of surgery.

David Henderson, Fildes's second in command at Porton Down, was a thirty-seven-year-old Scotsman who was as handsome and arrogant as the romantic hero of a Jane Austen novel. He loved animals and had gone to work on a farm after graduating from the University of Glasgow but had left in a huff because the farmer's methods were "unscientific." He returned to school and won a master's degree with a thesis on the diseases of sheep, and then a Ph.D. from London University with a study of anaerobes such as *Clostridium botulinum*, the cause of botulism.

His main contribution to the developing science of germ warfare, however, was to have invented the "Henderson apparatus," a mechanism for exposing small animals to aerosol sprays of disease-causing organisms. The apparatus, called a "piccolo" by the Porton Down scientists because of its appearance, was a tube about three feet long and two inches in diameter with holes along the side. The holes in the side were not for making musical notes but rather were "animal exposure ports," into which the heads of small lab animals were stuffed so as to give them a precise and measured dose of the pathogen.

The procedure was to place the test animal, usually a mouse, into an "animal holding cup," a small metal can, so that its body was inside the can while its head protruded from the open end. The experimenter attached the holding cup to the exposure port: the mouse's head stuck through the hole in the piccolo while the rest of its body remained outside. A rubber seal around the exposure port made for an airtight fitting at the attachment point.

Three or four canned animals could be fitted into as many exposure ports in the side of the apparatus, after which the experimenter sprayed an airborne mist of pathogen—the anthrax microbe or some other agent— through the interior of the Henderson tube. The animals would have no choice but to inhale a given quantity of the microbe. At the end, the experimenter detached the animal holding cups from the apparatus and released the animals into a cage.

Fildes and Henderson had run such tests again and again at Porton, in a small redbrick building at the edge of the campus, and before long hundreds of mice, rats, guinea pigs, and rabbits got their heads stuffed into Henderson tubes of different sizes for their allotted doses of airborne anthrax.

Fildes had sent reports of these experiments to George Merck, who had circulated them within the small circle of scientists and military men who were involved in the American germ warfare study. The Americans regarded the experiments as mildly interesting but didn't think they proved the feasibility of germ warfare. The battlefield, after all, was not a Henderson tube. What was needed were some real-life, open-air experiments with a biological agent such as anthrax disseminated over animals the size of humans.

Fildes and Henderson had come to see George Merck precisely for the purpose of telling him about a set of such experiments they had just recently concluded on a small island in northwest Scotland.

Even for the Scottish highlands, Gruinard Island was notably forsaken and remote. The island lay in a domain of rugged mountains, sudden storms, driving rain, mists, low scudding clouds, fog, wind, snow, ice, and gloom. The single road that bordered Gruinard Bay was more commonly overrun by sheep and wild goats than traveled by cars. The island itself, a 522-acre land mass, was across a stretch of choppy water and accessible only by boat.

Gruinard Island was kidney shaped, with the long axis running in a north-south direction. At the southern end a thin peninsula of smooth brown rocks—the British called it "shingle"—jutted out into Gruinard Bay. The spine of the island rose northward to two humps, the higher of which was a peak called An Eilid. On the summit, at an elevation of 350 feet, there was a large stone cairn. The land below was brown and marshy, covered with low brush, tan grasses, and deep wet peat.

The island was part of the Gruinard Estate, a large private holding belonging to Mrs. R. G. C. Maitland, wife of an Edinburgh advocate. Mrs. Maitland had purchased the estate in 1922, and at the beginning of World War II, Gruinard Island was her own private property. In 1881 six crofters had lived there in cottages built from the island's gray stone. They raised crops on part of the island and ran sheep and cattle on the rest. Once in a while, on sunny Sundays, a few mainlanders visited "Grunyard," as the natives called it, for afternoon picnics and egg collecting.

By 1941, a single shepherd remained on the island. He lived in a small stone hut on the southern end just above the shingle peninsula. There he tended some sixty sheep owned by a local butcher.

In December of that year, agents from the British Ministry of Defence arrived on Gruinard Island, cast out the shepherd and the sixty sheep, and declared the site a prohibited area. Later they formally requisitioned the island from Mrs. Maitland, paying her in compensation the grand sum of £500. Within the Ministry of Defence and at Porton Down, the island was henceforward known as "X Base."

In the spring of 1942, the British Pioneer Corps and Royal Engineers came to Gruinard Bay and built a camp for fifty enlisted men at

Mungasdale, an estate on the mainland across from the island. There they erected a couple of Nissen huts—corrugated steel buildings in the shape of half-cylinders—and ran a telephone line to the Royal Naval Base at Aultbea, the so-called Boom Defence Depot, about ten miles to the southwest. They converted one of the Mungasdale outbuildings into a laboratory and filled it with lab glassware, diagnostic reagents, instruments, and machinery.

On the island itself the workmen cleared and repaired the stone animal pens and fixed the limits of a decontamination area at the high-water line near the base of the shingle peninsula. Finally, a few hundred yards up the hill from the shepherd's hut, they erected a six-foot-high wooden gallows. Bombs filled with a suspension of the anthrax microbe, *Bacillus anthracis* (which the British had code-named "N"), would be suspended from the crossbeam and detonated, their vapors blown by the wind across a line of sheep. Sheep approximated the human body in weight, were highly susceptible to anthrax infection, and were plentiful in the area. In Scotland, indeed, there were more sheep than people.

The Western world's first successful biological weapons test occurred on Wednesday, July 15, 1942. The entire Porton Down Biological Department had assembled for the event at Gruinard Bay: Paul Fildes, David Henderson, Donald Woods, O. G. Sutton, and others. Sutton was head of the Planning and Reporting Section at Porton and it was he who had purchased a supply of sheep from the mainland farmers, and had them shorn, marked, and corralled at the Mungasdale camp until the day of the test.

The day began with a parade of men and animals to the Mungasdale jetty for the crossing over to Gruinard. A shepherd and his dogs, along with a couple of local children, herded fifteen white sheep in a pack toward the pier. There the Porton scientists plus a few helpers lifted the animals one by one onto a large, flat-bottomed rowboat. The men rowed themselves and the flock across the strait, a distance of about a half mile.

There was no pier on the Gruinard side, so the crew beached the boat on the east side of the shingle peninsula and lifted out the sheep. The shepherd and his two dogs herded the animals into the stone pen next to the shepherd's hut.

On the ground next to the stone pen were fifteen wooden crates, each just big enough to confine a single sheep without any extra space or wiggle room. At the front of the crate was a wooden panel with a half-moon cutout for the animal's neck; a separate panel with the corresponding top cutout slid down through a slot, immobilizing the animal like a thief in a pillory.

One at a time, the men picked up the sheep, placed them into the crates, and locked their heads in the front panels. Then, to prevent any of the anthrax spores from falling on the sheep's coats during or after the test, they covered the top and sides of each crate with a sheet of canvas. The scientists did not want the sheep ingesting any spores by licking them off their own wool; they wanted the results to occur by inhalation alone.

The experimenters loaded the fifteen covered crates and an equal number of air sampling devices onto two flatbed trailers stationed at a boundary line marked off by a token fence of post and string, the clean/dirty dividing line. The shingle peninsula, the shepherd's hut, and the decontamination area were on the "clean" or uncontaminated side of the line; everything higher up the hill than that, including the bomb gallows and the sheep grid area, was on the contaminated side. Workers on the clean side wore ordinary street clothes. Those across the fence were in protective suits consisting of heavy cloth coveralls, rubber boots and gloves, goggles, a gas mask, and a white cloth hood.

A Caterpillar tractor now started up and towed the flatbed trailers loaded with the crated sheep and the air samplers up the hill toward the test site. There the Gruinard spacemen unloaded the sheep and arranged them out in a ninety-yard arc a hundred yards downwind from the bomb gallows. The ground was wet and boggy and their feet sank into the mud as they walked.

Finally the crates had all been lowered to the ground and were all pointed in the same direction, into the wind and toward the bomb. Beside each crate was a glass vacuum device to sample the cloud as it passed over each sheep in succession.

At the gallows, David Henderson with no goggles, gloves, face mask, or any other protection, opened a glass flask of anthrax spore suspension and poured it into a thirty-pound bombshell, then repeated the process until the bomb case was full. Alan Younger, the explosives expert, hooked up the bomb's detonator to an electrical cable that ran to the control point, a small open space behind a bunker.

The men then left the test site and got out of range, on the windward side of the blast point. The wind was blowing at thirteen miles per hour, gusting to seventeen. The warning flag waved in the breeze.

And if this were a Hollywood movie, then in slow motion the bomb would drop from the gallows, it would fall to the ground inch by inch, while a high-pitched vocal chorus, something from *Carmina Burana* or *The Damnation of Faust*, wailed in the background, and then the bomb would strike the ground and slowly shatter into a million pieces, the anthrax suspension spraying up and out. The cloud of vapor would get caught in the wind and would bear down upon and envelop the line of boxed and waiting animals.

Which in fact is almost precisely what happened, except that the bomb never reached the ground but instead exploded in place at a height of four feet. And then, without sound effects of any kind other than the momentary clap of the detonation, a gust of anthrax flew on the wind across the line of sheep. In a matter of seconds the anthrax cloud passed over the animals, blew off toward Gruinard Bay, and then continued out into a stretch of open sea.

The Gruinard spacemen returned to the sheep and placed canvas hoods over the faces of the animals to prevent them from inhaling any stray anthrax spores. They lifted the fifteen boxed and hooded sheep up onto the flatbed trailers, and the tractor towed the trailers to the holding area, a grassy paddock high up at the edge of a cliff on the west side of the island. The scientists could see a few white houses on the shore across the water, the tiny village of Laide. At the holding area they released the animals from the crates and staked them to posts in long lines. Each sheep now had its own separate walking space, and no animal could touch the next.

Deaths began on the third day. When the Gruinard spacemen returned that morning, they found dead sheep lying on the grass, lines of dried blood running from their nose and mouth. Others were in their death spasms, while a few others remained healthy. Eventually, all but two of the fifteen animals died; only those at the extreme end of the line survived.

Finally, there were the postmortem examinations. To prove that the animals in fact died of anthrax and not some other random cause, the scientists took blood smears from each dead sheep and performed autopsies.

Like everything else at Gruinard, the autopsies were stark and primeval, done in an alcove along the cliffside where water ran out of the wall in a thin stream. One of the men placed a sheep on its back on a rock ledge, the open-air operating table. Then Reggie Bamford, the Royal Army Veterinary Corps surgeon, slit open the animal's belly from stem to stern. He cut out the liver, spleen, and some other organs, and dropped them into a galvanized pail.

The scientists were now finished with the animals. They flung their carcasses over the cliff and onto the rocks just above the water line. They threw peat-filled sandbags down on top of them to hold their bodies in place.

Then they set off a dynamite charge at the edge of the cliff. A shelf of rock flew up and out, slid down the cliff face, and buried the animals to a depth of ten feet.

The original test on July 15, Fildes and Henderson told Merck, was followed by another one on July 24 with fresh sheep placed along two lines, 100 and 250 yards downwind from the burst point. There had been a sudden wind shift as the bomb exploded and only half of the animals got doused with pathogen. Those that did expired precisely on schedule, however, which meant that the two trials taken together proved beyond any doubt that bacterial bombs worked.

All this was major news to George W. Merck and the Chemical Warfare Service officers, all of whom until recently had been in thrall to the Leon A. Fox view that bacterial bombs couldn't work because the agent inside them would be killed by the heat and shock of the explosion. Fox had been correct up to a point, for the biological agent in the Gruinard bombs had indeed been killed in large numbers: perhaps ninety percent of the anthrax spores had been destroyed in the explosions.

But the amount of pathogen *killed* by the explosion didn't matter; it was the amount that *survived* that was important. If enough biological agent survived to kill the target, then it was irrelevant how much of the agent had died. The ten percent that survived on Gruinard was more than enough to kill every last sheep the cloud had reached, and if the exposed sheep had been enemy soldiers, many if not all of them would now be dead.

The British had also performed a second type of experiment on Gruinard Island. They hollowed out a supply of twenty-millimeter bullets, filled them with anthrax spores, and then fired them through armor plating into a closed cubical tank containing live sheep.

The sheep died, although not necessarily from being hit by the bullet. It was enough for the sheep to breathe the atmosphere created by the bullet's entering the chamber and releasing its spores. These anthrax-filled bullets could be useful, Fildes suggested, as antitank weapons: not only would they kill the tank personnel, but the vehicles would probably remain infected ever afterward, making them unusable by fresh crews.

The Gruinard test series had been capped by an actual aircraft drop when on September 26, 1942, a two-engine Vickers Wellington bomber flew over the island at an altitude of 7,000 feet and released a thirty-pound anthrax bomb into the year's final layout of animals. This was to be the crowning moment of the biological warfare trials at X Base, and for this occasion the scientists had set out a record fifty sheep.

Not a single one of them died. The bomb sank into a peat bog and discharged its anthrax filling into the ground instead of the air.

The scientists redeemed themselves a month later when they repeated the airdrop on a hard sand beach at Penclawdd on the coast of Wales. The device hit the hard ground, the anthrax slurry burst out in a fine spray, and the sheep died in a timely manner. The message to George Merck and the Chemical Warfare Service officers was clear enough. Germ warfare was no longer science fiction or a lurid Sunday-supplement mirage. Biological bombs could be, and were, practical and deadly weapons.

That being so, said Fildes, he wanted to ask the Americans for their help. While the British were arguably ahead in both the theory and practice of biological warfare, the Americans were far better equipped than they, in terms of manpower and physical plant, to churn out vast masses of offensive microbes, which was their primary and urgent need at the moment.

Specifically, Fildes requested that the Americans undertake to build fermentation plants for the manufacture of large quantities of agent N, anthrax, and for similarly plentiful supplies of agent X, botulinum toxin, the most deadly substance known to man. In fact, they wanted approximately three kilos, or seven pounds, of dried botulinum toxin at the earliest opportunity.

Fildes even put it in writing, exactly as if he were placing an order: a formal request for "three kilo dried X." He gave the sheet of paper to George Merck. Then he and Henderson said their formal good-byes to Merck and the others, and they left.

3

wo weeks after Paul Fildes placed his botulinum order with George Merck, Ira Baldwin got The Call.

Ira Baldwin, at that point, was an agricultural bacteriologist and chairman of the bacteriology department at the University of Wisconsin, where he was doing peaceful research on arcane and academic subjects like the role of root nodule bacteria in agriculture. He was short and dumpy, wore glasses, always had on a three-piece suit, and had thinning and grayish hair at the age of forty-seven. He was kind and gentlemanly and the last person anyone would imagine as the scientific director of an American biological warfare research center.

But one day in late November he got a telephone call from Colonel William C. Kabrich, chief of the Technical Division of the U.S. Army's Chemical Warfare Service at Edgewood Arsenal in Maryland. Kabrich was calling to invite Baldwin to a meeting at the National Academy of Sciences in early December. It would be a small gathering, he said, just Kabrich, Baldwin, and a few other scientists, plus a few military men, concerning a subject of national importance that he couldn't discuss over the phone.

Well, it was wartime, so Ira Baldwin accepted.

The academy was housed in a long three-story building that occupied a whole block on Constitution Avenue in Washington. It looked out on

Potomac Park, the Lincoln Memorial to the right and the Washington Monument off to the left. As a setting for a scientific conference, it was impressive.

As was the group in the meeting room. Half of them were in uniform, and Baldwin recognized most of the rest, many of whom were bacteriologists like himself. There was Jim Sherman from Cornell; René Dubos from Harvard; Paul Hudson from Ohio State; and E. B. Fred, dean of the graduate school at Baldwin's own home base, the University of Wisconsin. Dr. Fred, Baldwin remembered, had been called to Washington about a year earlier to head up the WBC Committee, the War Bureau of Consultants, whatever that was. All of it had been extremely hush-hush at the time.

Colonel Kabrich began the meeting with a secrecy message of his own, talking about forty years in prison and a fine of $10,000 if any of what he was about to say ever got out to a living soul. He seemed serious.

The fact was, he said, that according to intelligence reports both Japan and Germany were engaged in preparations for germ warfare. World War I had been the dawn of gas warfare, but however successful gas warfare might have been, science had now progressed beyond that, to the point where live bacteria could be used in place of dead chemicals.

Bacteria were even more insidious than gas. They were invisible and odorless. They could crawl inside gas masks, penetrate clothing, invade the body through the skin—or even in the very act of breathing. The enemy could conceivably wipe out a battalion—or a whole city, for that matter—by dropping bombs loaded with microbes against which there were no vaccines, cures, or any other defenses.

There had been a time in the past when the Army had been convinced that the use of bacteria wouldn't work, but the British had done tests that indicated quite the contrary, that bacterial clouds were highly efficient killing mechanisms. What the British tests did not establish, and what was currently unknown, was whether pathogenic bacterial agents could be mass-produced, and above all whether they could be mass-produced safely.

"Would it be possible to produce tons, and I mean *tons*, of living pathogenic microorganisms?" he asked. "Could you do it so that they'd retain their virulence? And could you do it safely, both for the workers and for the surrounding community?"

The academic microbiologists around the table mostly raised their eyebrows at this. In the laboratory, bacteria came in extremely small quantities. You had to culture the stuff painstakingly in petri dishes, or in test tubes or Erlenmeyer flasks, in order to create any appreciable amount of a given microbe. A shot-glass quantity was typically the largest volume you ever dealt with, an amount measured in grams or ounces. You never made even so much as pounds of microorganisms, much less "tons."

But to Ira Baldwin, there was no reason in principle why you couldn't. "Well, if you could do it in a test tube, you could do it in a 10,000-gallon tank," he said. "I don't know how many 10,000-gallon tanks it will take to produce tons, but if you get enough tanks I'm sure you will get tons." He said furthermore that if you did it correctly you could do it more safely on a batch-production basis in mass volumes than you could by working in a conventional laboratory with test tubes. There seemed to be no doubt whatsoever in his mind that this could be done.

And that was that. Baldwin went home feeling that by venturing this concise opinion he'd done his part for his country's war effort and could now get back to peaceful and normal research pursuits.

But no. Ten days later, Kabrich was back on the phone. He was now Brigadier General Kabrich: promotions came fast in wartime, he said. He wanted Baldwin to return east on December 22.

"What for? That's getting awfully close to Christmas," Baldwin said. "How long do you want me to stay?"

"For the duration," said Kabrich. "We want you to come out here and show us how to do what you told us could be done. We want you to go ahead and do it."

Although he did not say it in so many words on the phone, Kabrich was clearly asking Baldwin to come east and start growing "tons" of killer bacteria. And not for humane uses in medical research, or to produce vaccines, but rather for the ultimate purpose of killing people.

Baldwin had some thinking to do, for he was not a born warrior. He'd grown up on a forty-acre farm in Indiana where during the summers he would with his own bare hands husk a hundred bushels of corn per day. To make money for college he raised and sold flocks of Indian Runner ducks— peaceful white creatures that gathered around him when he brought the feed pan out from the barn. His grandparents were Quakers, and although he wasn't a Quaker himself he was strongly religious and had even preached in small country churches where there was no regular minister. But he wound up a man of science, getting degrees in agricultural chemistry and bacteriology, and he had just been installed as chairman of the bacteriology department at Wisconsin when the call had come from Washington.

Which meant that he now had to wrestle with the considerable moral problem of using microbes to kill people, instead of killing the microbes to save people. This would require a complete shift in moral perspective. Still, it only took him twenty-four hours to think his way through the problem and make the necessary mental alteration.

After all, the immorality of war was war itself, he thought. It was the very act of killing that was wrong, not so much any specific method of killing as opposed to some other one. He tried to put himself in the position of being killed in battle. Would he rather be maimed by high explosives, dismembered, burned to death by fire and flame, ripped open by a bayonet, his ruptured guts spilling out in front of him . . . or would he rather die of the worst disease he could possibly think of?

He thought he would rather have the disease.

He imagined himself in a hospital. He'd been to hospitals, God knew. His youngest daughter, Frances Mary, had been killed in a car accident at the age of sixteen, just the year before, in 1941. She and five classmates had been coming home from an afternoon party. They were walking across an icy bridge, a car skidded on the ice, it slammed into them, and three of them were dead and three were mangled beyond belief.

And so he gained firsthand experience with accident wards. In an accident ward everybody who wasn't knocked out by drugs was in intolerable pain. Their bones had been broken, their muscles had been crushed and torn, and you could hear them moaning, you could see their bodies twitching involuntarily. An accident ward was the last place in the world any sane person wanted to be.

But infectious disease wards weren't quite the same; they didn't harbor suffering of the same order of magnitude. Ira Baldwin had had personal experience with those, too, back in World War I when he was a second lieutenant on the Field Artillery Range at Camp Taylor, Kentucky. This was in 1918, during the great flu epidemic, the worldwide scourge that had killed some twenty million people, more than were killed in the war itself.

Plenty of soldiers had died at Camp Taylor from the flu. They died not of bullets but of germs, and Ira Baldwin had been in charge of the burial detail. There was one burial detail per day, and he and his men would go into the influenza ward and collect the newly dead. It had been bad in there, but still the patients were not crying out in pain. They were sick, they were weak, they were dying, but all the same they didn't seem to be suffering sheer physical pain.

Diseases, by and large, didn't bring on suffering, Baldwin decided. It was physical injury, the body being penetrated by knives or bullets, burned by fire, or physically torn asunder by mechanical forces, that brought on the severest pain and agony.

So it turned out to be a simple decision in the end. If it was a question of how much people suffered, then germs and diseases, bad as they were, were less bad than bullets, flames, or shrapnel from hand grenades or iron bombs. Barbaric as it was to be killing people at all, killing them with diseases was more humane than killing them with the usual stuff.

Ira Baldwin arrived at Edgewood Arsenal on the night of December 21, 1942, four days before Christmas. Edgewood was the headquarters of the U.S. Army chemical works, the place where, during World War I, the United States had produced a vast tonnage of offensive chemicals: 935 tons of phosgene and 711 tons of mustard gas, among other things.

When Baldwin got off the train there was fresh snow on the ground. A soldier by the name of Schwanke picked him up at the station and drove him to his new home, which turned out to be the guest house on the base's golf course.

The first day at work, Baldwin found that he'd be sharing an office with General Kabrich. Kabrich was an excitable sort, with quite a temper, and during his perpetual meetings and conferences he seemed to bounce from the chair to the ceiling and from side to side. This was somewhat unnerving to Baldwin, as it was hard to concentrate properly in Kabrich's presence. But soon enough he was busy recruiting a staff from the ranks of his former students and colleagues back at the University of Wisconsin.

Then he began his search for the site where he was to cultivate lethal germs by the ton. It had to be close to Washington, but not *in* Washington, and it had to be remote but not *too* remote; such were his instructions from General Kabrich.

A couple of Army men—James Defandorf, chief of the Army's medical research division, and Arvo T. Thompson, of the veterinary corps—had already made a survey of possible sites for a biological warfare research center, but they hadn't liked a single one. Rolla E. Dyer, director of the National Institute of Health in Bethesda, Maryland, had offered them some space on the fourth floor of one of their laboratory buildings. Defandorf had considered this for a while but rejected the idea: there would be no place on the campus for field trials. Besides, he was concerned about the possible repercussions of devoting part of a U.S. Public Health Service lab to offensive weapons research.

Defandorf had then turned his thoughts toward the remote and apparently deserted islands of Chesapeake Bay. They were self-contained, isolated, and perfect for top secret war work, so in July of 1942 he and Arvo Thompson, together with some others from the Chemical Warfare Service, paid a visit to the Chesapeake Biological Laboratory on Maryland's Solomons Island. The biological lab itself, which did marine research, was not available for his purposes, but the director, Dr. R. V. Truitt, suggested several other locations in the Chesapeake: Barren Island, Taylors Island, and Wroten Island, among others. In fact, he offered to take them out on his motor boat, the *Mahatru*, for a personal inspection tour of the sites.

Barren Island, however, did not live up to its name, there being a hunting club on the premises. The other islands were either too large or too small, too inaccessible, or had some other disqualifying feature.

In August, Defandorf and Thompson visited Mt. Weather in Virginia, site of a U.S. Weather Bureau monitoring station. It was suitably remote, but high in the mountains, where in the winters the snows were often deep.

They visited Sugarloaf Mountain in Frederick County, Maryland. The entire place was privately owned—the whole mountain—but there was not enough flat and open space there, and the road to the top was narrow and winding.

At this point, Ira Baldwin entered the picture and undertook a survey of his own. He visited the Bata Shoe Factory, a group of vacant buildings just outside of Aberdeen Proving Ground, a few miles away from Edgewood. Not quite what they needed, a shoe factory.

He was offered the grounds of a state teacher's college somewhere in Alabama. No.

Then one cold day in February he drove out to Frederick, Maryland, to the site of an abandoned National Guard flying strip. The motor pool car had no heater, and the trip took about three hours, so Baldwin wasn't in the most receptive frame of mind by the time he arrived at Detrick Field.

All the planes that had been based there had long since been flown to Europe to fight the new world war. What remained was a large empty hangar, a concrete-block control tower, a runway, and some two-story wooden barracks. Behind them, receding into the gray haze, lay a pleasant stretch of fields and pastures.

Lots of potential here, Baldwin thought to himself. Room enough for any number of pilot plants and laboratories, and plenty of space for field tests. Off in the distance was a range of low mountains, and the whole scene was pretty as a picture.

By the time he got back to the car, a light snow had started to fall—at which point he discovered that the car's single windshield wiper didn't work.

No matter. This was the place. Here was where they'd build the labs and the fermenters, the 10,000-gallon tanks that would produce the Army's deadly germs.

All kinds of germs, as many as they wanted.

Tons.

By the time Ira Baldwin had found a site for the American biological warfare works, Shiro Ishii and his men had moved into a new facility of their own at Ping Fan, some twenty miles south of Harbin, China.

Ping Fan was a walled city covering more than two square miles of ground and incorporating, at the end, more than 150 individual buildings and structures. There were laboratories, dormitories, munitions dumps, barns, stables, dissection and autopsy rooms, greenhouses, a farm, a prison, a power plant, and incinerators, plus, for the Japanese workers, a library, a bar, restaurants, gardens, and recreational facilities that included athletic fields, swimming pools, a thousand-seat auditorium, and, for the convenience of its hardworking male crew, an on-site brothel. The perimeter was fortified by watchtowers, a moat, and a fifteen-foot-tall brick wall topped by high-voltage lines and barbed wire. The nature of the enterprise was camouflaged by the two names "Anti-Epidemic Water Supply and Purification Bureau" and "Unit 731."

To the Japanese workers who arrived there from their farms and rural villages in the homeland, the size and scope of the place were remarkable. "When I set foot on this land which was drenched in the sunlight of spring, I felt as one who had just awakened from a dream and was gazing upon the dazzling light of the grand scenery that lay before my eyes," one of them said years later. "That brilliance was not due to the sun. It was the sight of rows of modern buildings looming unexpectedly in the middle of a vast plain.

"Firstly, the central buildings towering skyward over other buildings in the area, with all square-tiled facades, were larger than any of those I had observed on my trip over, including Osaka, Hsingkiang, and Harbin. These buildings reflecting the sunlight glistened in brilliant white and broke into the vast sky. High earth walls were constructed with barbed wire fencing atop. It was obvious that this compound was isolated strictly from the outside world."

Ishii and his men had moved into the first of these halcyon structures during the fall of 1938. Still, the place was so immense that it was not finished for two more years. By 1940, finally, some 3,000 personnel lived and worked on the premises. At the completion ceremonies Ishii lectured his medical staff about their solemn duty to God and country: "Our god-given mission as doctors is to challenge all varieties of disease-causing microorganisms," he said. "To block all roads of intrusion into the human body; to annihilate all foreign matter resident in our bodies; and to devise the most expeditious treatment possible.

"However, the research work upon which we are now about to embark is the complete opposite of these principles, and may cause you some anguish as doctors. Nevertheless, I beseech you to pursue this research, based on the dual thrill of a scientist to exert efforts to probing for the truth in natural science and research into, and discovery of, the unknown world, and, as a military person, to successfully build a powerful military weapon against the enemy."

They had ample resources at Ping Fan to perfect such weapons. Ishii and his men investigated the causative agents of plague, cholera, anthrax, glanders, dysentery, typhoid, tetanus, and tuberculosis, among other diseases. Their equipment for mass-producing these agents was, Ishii said, "inferior to none." The bacteria culture plant had four 1-ton boilers for preparing the bacterial culture medium and fourteen autoclaves, each of which could hold thirty cultivators for growing pathogens. At peak production, Ping Fan's bacteria culture system was capable of turning out 300 kilograms of plague organisms, 500 kilograms of anthrax spores, or as much as 1,000 kilograms of cholera bacteria per month. There was an entire section devoted to the creation of new types of bombs and a large room for the mass production of fleas.

The biological testing facilities were also quite comprehensive. Ping Fan had its own airport and a separate area into which planes could drop biological bombs. In all, Shiro Ishii regarded the complex as the world's largest and finest germ warfare facility.

■   ■   ■

Detrick Field, before Ira Baldwin got there, had been the training center for the 104th Aero Squadron of the Maryland National Guard. The site, a ninety-acre grass strip northwest of the city of Frederick, had been named in honor of Frederick L. Detrick, M.D., the unit's flight surgeon.

The 104th Aero Squadron had used the field for its annual two-week summer encampments, during which neat rows of canvas tents appeared on the grass to accommodate the pilots who flew in on their DeHaviland O-38 and Curtiss JN-4 "Jenny" biplanes. They'd take off in the early mornings, float up over the Catoctin ridge to the west, and engage in mock dogfights high above the valleys and cow pastures. Soon the squadron built a cinder-block control tower, dining hall, and latrine.

In 1940, with Europe into its second great war, the federal government leased the field for use in its Cadet Pilot Training Program. The Army erected the barracks and the large hangar, and poured concrete for an aircraft tie-down ramp, a taxiway, and sidewalks. That was the physical state of the place when on March 9, 1943, the Army Chemical Warfare Service took formal possession of Detrick Field, annexed some of the adjoining farmland for field trials, and renamed it all Camp Detrick.

Personnel started arriving in April. They were not pilots or even, for the most part, military men. They were civilian biologists whose task was to mass-produce germ weapons. Their first order of business was to fill the British production order for "three kilo dried X."

For a working biologist, this was a highly novel request. There was such a thing as mass production in microbiology, but it was concerned with making commercial quantities of benign and useful substances like baker's yeast, the manufacture of which Ira Baldwin had had some past experience with. In fact it was precisely his extensive knowledge of industrial fermentation techniques that had motivated the Army to select him as head of the nation's germ warfare project. But there was all the difference in the world between the factory manufacture of commercial yeast and the large-scale cultivation of botulinum, the least speck of which could kill a person.

The British, however, had already done much of the necessary spadework, recording their efforts in the "Green Book." The "Green Book" was the bible of the BDP, the Biology Department, Porton. It was a tight ream of British-size typing paper gathered and bound between two stiff green cardboard covers. The text, which was fleshed out with charts, tables, diagrams, maps, and photographs, was a concise distillation of everything the Porton scientists had so far learned about the art and science of biological warfare.

In May 1943, when Camp Detrick was just two months old, Paul Fildes sent a copy of the "Green Book" to Ira Baldwin. When Baldwin turned to the "P" section—"P" for Production, the mass cultivation of biological agents—he found that the anthrax spores used in the Gruinard tests hadn't been grown in small lots such as in test tubes or Erlenmeyer flasks. But then again they hadn't been grown in industrial-scale vats, either. The British had produced their spores in, of all things, milk churns, stainless steel milk cans of the type used for making butter. Each milk can had a capacity of fifty liters—thirteen gallons—roughly the capacity of an automobile gas tank, and evidently they had worked well enough as anthrax growth vessels.

Baldwin looked at the photographs of the British milk churns. There were four of them ganged together in a line, each connected to the next with flexible tubing. All the cans had stainless steel lids that were held in place by metal clamps. The first can in the sequence was for media preparation. The last can was for disinfecting the slurry at the end and was equipped with a vertical pipe as an exhaust flue. The two middle cans were the actual growth vessels, and they had spigots in the bottom for drawing off the contents: turn the tap and out poured a slurry of fresh-grown anthrax bacteria.

That, apparently, had been their entire production works.

Obviously it had been enough. Bacterial culture, after all, was not inherently complicated; it occurred all the time in nature, automatically. All you needed, basically, was a starter culture, a growth medium, and a container. And if what you were growing was *Bacillus anthracis*, the anthrax microbe, then you'd also have to pump air through the broth in order to get the bacteria to sporulate, for *Bacillus anthracis* was an aerobic bacterium and required air to reproduce.

The anthrax growth medium, Baldwin saw from the text, had been a typically British mixture of marmite (a commercial brewer's yeast), West Indian molasses, and distilled water, plus a few chemical herbs and spices. The growth cycle had taken some thirty-six hours from start to finish. They put a primer culture into the two culture churns, bubbled air through perforated aeration heads at the bottom, and let it sit for the next day or so. At the end of it they had simply drained off the liquid from the spigots. The final yield of a single growth cycle was a one-liter glass bottle of anthrax suspension.

Ira Baldwin had to admire the simplicity and ingeniousness of the Porton Down milk-churn system. Of course, it would have to be scaled up somewhat.

The initial production plant at Camp Detrick was called "Black Maria." That was the formal and official name of the place, the way it would be referred to by the staff ever afterward, and the title it went by in Camp Detrick special reports.

It was a two-story black tar-papered shack that looked like a chicken coop and gave every outward appearance of having been thrown together overnight by a couple of farmers. In reality, the H. K. Ferguson Company, industrial engineers and builders of Cleveland and New York, had come to Frederick in the late spring of 1943 and had put up the building in a couple of weeks.

The sole assignment of "Black Maria" was to produce the "three kilo dried X," the botulinum toxin that had been ordered by the British. That done, the building was to be demolished without a trace. Which was exactly what happened. "Black Maria" went up in a flash in May 1943, and by October of the same year it was gone. The building's mean appearance and fleeting life span gave "Black Maria" legendary status at Camp Detrick, as if it had existed only in myth. But the place was real enough.

"Black Maria" was enclosed within two nested levels of security, the first being an outer barbed-wire fence that separated the barracks and administration areas from the laboratories and pilot plants, all of which were then under construction. Inside the barbed wire, off in its own private space, was a nine-foot-high wooden fence that enclosed a small rectangular lot. Standing in the middle of it was "Black Maria."

Electrical lines stretched over the wooden fence to a corner of the building, and at night four large floodlights on the flat rooftop shone out in all directions. Night or day, soldiers equipped with machine guns stood watch in guard towers located diagonally across from each other just outside the fence.

Inside "Black Maria," on the top floor, was the British milk-can system scaled up by a factor of eight. Under three high windows on the east side of the building were four 100-gallon reactor tanks, each connected to the next by valves and piping. Starter culture went in at the top, the new growth came off at the bottom, forty-eight hours from start to finish. The major difference from the British system was the lack of aeration heads inside the 100-gallon tanks. The tanks needed no aeration because they would be growing the botulinum microbe, which was anaerobic and proliferated only in the absence of oxygen.

Baldwin and his crew made a few improvements, of course, chief of which was the steam-sterilization system. In any bacterial mass-production operation, the pitfall to avoid was cultivating the wrong microbe by mistake. Nature was chock full of microbes: they were in the air, in the water, on surfaces, in your mouth, on your hands, everywhere imaginable. They were also inside the fermenter tank, which was, after all, just a large inner surface and a volume of air, both of which could be harboring vast quantities of different contaminants. The greatest embarrassment for a Camp Detrick bacteriologist was to drain the fermenter tank and suddenly discover that he'd grown not the target microbe but some other one, or maybe even a whole slew of competing bacteria, all because he hadn't adequately sterilized things beforehand.

The way around that was to kill the other bacteria before you started, and at Camp Detrick the disinfectant of choice was steam. Each of the 100-gallon fermenter tanks was therefore equipped with an external water jacket through which hot scalding steam could be forced under pressure. The tank's internal surfaces would be sterilized by the heat, as would the growth media itself.

The steam would be provided by a wood-burning furnace and a steam boiler located in a small hut just outside the wooden fence surrounding "Black Maria." After the heat treatment, the tanks and their contents would be cooled off by cold water circulating through the same external water jacket.

Other Detrick improvements included the addition of paddles inside the fermenter tanks with which the mixture could be stirred before the start of the growth process and a closed plumbing system that allowed the tanks to be drained directly into settling tanks on the first floor.

In June 1943, a crew of four scientists, all of them Ph.D.'s in either chemistry or bacteriology, assembled in "Black Maria" and began work. They were Alvin Pappenheimer, John Schwab, Mike Foster, and Bill Dorrell. Pappenheimer, leader of the group, was a Harvard University toxicologist and a world expert in *Clostridium botulinum*, the botulinum microbe. He was in fact the first person ever to be immunized against the toxin.

Earlier that year in his lab at the Harvard Medical School, Pappenheimer and his colleagues had worked with a version of *C. botulinum* known as the Hall #57 colony, a strain that proved to be highly toxic to mammals. In animal trials, the Hall #57 strain had killed all test mice within seventeen hours of inoculation. That would be the strain they would try to mass-produce in the 100-gallon fermenter tanks.

To optimize the growth process the Harvard researchers had also experimented with thirty different types of culture media in the lab before deciding on one whose major ingredient was corn steep liquor, an extract of pressed corn kernels.

The production cycle in "Black Maria" began with the scientists sending steam through the water jackets of the empty 100-gallon fermenter tanks, then filling the tanks with the corn steep liquor and heating it almost to the boiling point. That sterilized the medium. After a prescribed time, they sent cold water through the jackets, cooling it down again. Then they were ready to inoculate the tank.

There was a six-inch-wide hand hole in the top of each fermenter, and they removed the stopper, poured in a flask of the prepared starter culture, and replaced the stopper. They turned the hand cranks on the top of each fermenter, and the agitator paddles blended the mixture into a smooth broth. By that point the bacterial cells were already replicating themselves and releasing their toxin into the surrounding medium. For the next two days, the tanks simply sat there quietly while the bacteria inside them propagated, multiplied, and exuded toxin. At the end, if all went well, each tank held a hundred gallons' worth of extremely dilute botulinum.

The researchers drained the slurry into the settling tanks on the ground floor of "Black Maria" and then added a coagulating agent to the mixture. The toxin molecules clotted together into bigger clumps that settled out at the bottom of the tank. The men then transferred the consolidated mix to a second and final settling tank.

The final settling tank was long and slender, with a conical base that terminated in a short vertical pipe, the outflow from which was controlled by a valve. The scientists connected a five-gallon glass bottle to the outflow pipe, and at the end of the settling period they opened the valve and drew off the doubly-concentrated toxin.

Still, they had to establish definitively that it was "bot toxin" they'd grown in the tanks and not some harmless contaminant microbe. Over in a corner of "Black Maria" were some stacks of animal cages filled with mice, and the scientists now injected a sample of the liquid into half a dozen or so of the animals. If the mice remained healthy, no toxin had been grown. Mass death, by contrast, meant good-quality toxin, and the researchers often went through a dozen mice per day.

Off in another corner was an autoclave, a device for high-pressure steam sterilization. The experimenters placed the dead mice in the autoclave and sterilized them to kill any remaining traces of pathogen. Finally Alex Bryant, the Army private whose job it was to make steam in the boilers, incinerated the animal carcasses in the wood-burning stove, after which he buried the ashes.

The four scientists worked around the clock in eight-hour shifts, seven days a week, and after about two months of initial safety testing, trial runs with mock agent, and then actual production runs with the hot agent itself, they had filled the British order.

4

While the "Black Maria" crew was manufacturing botulinum for the British, four other Americans landed in England for a program of practical, hands-on experience, or what might loosely be called biological warfare training. They were William B. Sarles, Calderon Howe, and Harold P. Carlisle, all of whom were U.S. Navy medics, plus Carl E. Venzke, a U.S. Army veterinarian.

As soon as they set foot in London it was clear to all of them that there was a war on. They heard the sound of an air-raid siren, a warning to take cover, but Venzke, who had never been out of Iowa, went up to a rooftop to watch the action. He heard a hum coming closer, a buzz bomb. It landed and exploded nearby, taking with it part of a building. This was definitely not like being in Iowa.

The four of them went by train to Salisbury, the nearest city to Porton Down, where they were put up at the Old George Hotel, a half-timbered Tudor inn on High Street. The hotel was far older than the United States itself, having gone up in the year 1320. Salisbury Cathedral, an immense gray stone structure just two blocks away, was even older, having been built between 1220 and 1258.

The Old George Hotel was about eight miles from Porton Down, and all of them, together with Paul Fildes and David Henderson, who were also living at the hotel, commuted in every day by bus.

The biological division at Porton Down was in a low-slung redbrick structure that housed their agent production works, animal exposure chambers, storage rooms, and offices. In contrast to the war going on in London, everything here seemed to be quite calm and civilized. The workday started at nine and ended at five and Fildes served daily tea in his office at three o'clock.

A short distance from Fildes's office was Building 230, also known as the "Bun Factory," where between October 1942 and April 1943, workers had mass-produced five million units of the United Kingdom's first operational biological weapon, the anthrax-filled cattle cake. The cattle cakes, which looked like small dog biscuits, were made of ground linseed meal, and there was a small depression in the center of each "bun" for a tiny dose of anthrax spores. The plan was to fly over Germany in bombers and drop the spiked buns onto the area's prime cattle-grazing pastures. The animals would eat the cakes, contract anthrax, and die a quick death. The British referred to the scheme as "Operation Vegetarian."

Anthrax was an ancient disease, recognized as a distinct illness as far back as the Greeks. It was most common in sheep, cattle, pigs, horses, and goats, who contracted it by eating grass or drinking water that contained the microbe. The animal developed a fever and then a bloody discharge from the nose and mouth. It staggered and swayed, then its legs gave way, and it collapsed on the ground. Sometimes there were convulsions, the animal's limbs twitching uncontrollably. Then, finally, it died. The microbe was fast-working, and in acute cases death occurred a few minutes after the appearance of the first symptoms.

Anthrax was easily transmissible to humans, among whom the results were equally bad, if not worse, depending on the mode of transmission. In cutaneous anthrax, where the microbe entered the body through the skin, there was an incubation period of from one to five days after which the victim developed a skin rash at the site of entry, and ugly black scabs appeared on the skin. Cutaneous anthrax was treatable, and if it was recognized in time, the patient recovered.

Then there was gastrointestinal anthrax, contracted by eating infected meat that had not been fully cooked. The disease showed up as severe abdominal pain, fever, vomiting, and bloody diarrhea. Without treatment, death followed in about fifty percent of cases.

The worst form of the disease was inhalational, or pulmonary, anthrax, contracted by breathing the microbe from the fur or hides of infected animals. Infection by inhalation was known as wool-sorter's disease, as it occurred among those who handled contaminated sheep's wool.

This was the most deadly form of anthrax, for by the time the first symptoms appeared it was already too late for treatment, and mortality rates were in the ninety-five percent range, whether treated or not. The microbe entered the lung, where it germinated and multiplied, and from which it then swarmed through the bloodstream and exuded toxins while the victim experienced mild chest pains, malaise, cough, and fever, the symptoms of a common cold. The symptoms lasted for a day or two, after which there might be a short period of apparent improvement. Then, usually within twenty-four hours of the apparent remission, the victim died of respiratory failure.

Bacillus anthracis, in addition to being fast-acting and lethal, had a third property that made it ideal for use in a biological weapon, and that was its ability to sporulate. When faced with adverse conditions such as lack of proper nutrients or moisture, certain types of bacterial organisms compacted themselves up into little balls and extruded around their outer surfaces a tough and durable protein coat. In that state, the organism was known as a spore. Spores were impervious to light, heat, and radiation, and even to some noxious chemicals. They could last that way, ready to germinate when conditions were right, for decades, and perhaps even for hundreds of years.

Sporulation, in other words, was God's gift to germ warfare. It was precisely the indestructibility of the anthrax spore that had allowed enough of them to survive the detonation of the biological bombs that had killed the sheep on Gruinard Island. Those devices, however, had been large thirty-pound canisters, whereas the British had a smaller and perhaps more efficient bomb in the inventory that they could mass-produce for use as a biological weapon: the four-pound "Type F" munition.

The British had designed the Type F bomb as an incendiary device to ignite wooden buildings, but the canister could be converted to hold a biological filling. The bomb was a welded steel tube some twenty-one inches long and about an inch and three quarters in diameter, closed at one end by a metal cap and at the other end by an explosive fuse. A thin shaft of high explosive ran down the center of the tube, the so-called axial burster, and the liquid biological agent was poured in around it. The cavity held about a pint of liquid slurry; on detonation, the axial burster would blow open the steel walls of the bomb, at the same time aerosolizing the anthrax spores and dispersing them as a homogeneous cloud.

That, anyway, was the theory. The actual suitability of the four-pound device as a biological bomb would have to be demonstrated by tests, with live anthrax agent and live animals. And so in August 1943, this time accompanied by the four Americans, the British returned to the Holy Land of Biological Warfare, Gruinard Island.

There would be some changes in procedure from the previous year's tests. In 1942, when the dynamited rock slid down the cliffside, the force of the rock fall had dislodged one of the sheep from its final resting place. The sheep, with a halter still on it, dropped into the water and floated over to the mainland.

A local dog found the carcass and started feeding on it, and shortly thereafter the dog had contracted anthrax. It recovered, but not before it had somehow passed on the infection to a few sheep in the area, and a total of twenty-five animals eventually died on the farms and estates bordering Gruinard Bay.

The British government came up with a story that a passing Greek freighter had dropped a diseased carcass into the ocean, and that this was responsible for the anthrax outbreak. British agents set up a payment center in the village of Aultbea and, "on behalf of the Greek government," compensated the affected farmers.

In the new wave of trials, therefore, the dead sheep would not be buried. They would be incinerated on the spot and their *ashes* would be buried.

In their off time while waiting for the weather to improve, the Americans sometimes hiked up into the mountains surrounding Gruinard Bay. They'd heard rumors of a solitary couple living on the shore of a lake together with their two small children who had smallpox, and one day Calderon Howe and Carl Venzke hiked up to the lake and located the family. When they arrived, the father, a sheep herder, was holding a shepherd's crook, as if all of them had been transported back to biblical times.

The two children actually did have smallpox, two of the last cases of endemic smallpox in the United Kingdom. The disease was not curable and there was nothing the Americans could do. After a while they hiked back to the camp.

At last the weather was right and the time had come to do the tests. They put the animals in crates, set them out on the hill near the gallows, and then got back out of range and watched the bomb go off. Then they tied the animals to wooden stakes on the high cliff, left the island, and waited for results.

It was Venzke's job to monitor the sheep and take their temperature on each successive morning following the test. He'd show up there on the cliff in his space suit and gaze down on the staked animals. At first they seemed healthy enough and none of them was running a fever.

But then on day two or three he picked up a fever on a few of the sheep. He stood there and watched as they swayed back and forth and hobbled around unsteadily on the grass. Eventually the diseased animals collapsed on the ground and panted for air. Then they just faded away and died.

Venzke then postmortemed the animals, opening them up and taking a blood sample from the heart and tissue samples from the liver and spleen. He had no inclination for the task—it was not why he became a veterinarian—but there was no getting around the need for it.

One time, back at Porton Down, there had been a little display of poetic justice when Paul Fildes had taken one of the lab animals, a rhesus monkey, out of its cage to show it off to the Americans. The monkey jumped on Fildes's head, defecated on it, and rubbed the stuff around his bald pate.

That was as much revenge as the animals ever got.

The Americans at Camp Detrick were not quite so advanced as their British brethren in the matter of biological bomb testing. For one thing they lacked a suitable bomb. For a second thing they lacked a supply of live agent to fill it with. And last of all they lacked a suitable area for field trials.

But in July 1943, for the sake of getting started, the members of Camp Detrick's new M (Munitions) Branch did what they could. They got hold of a few unused chemical bomb casings from Edgewood Arsenal, filled them with mock agent, hauled them out onto the cow pasture at the back of the base, and set them off.

The bomb they chose for this exercise was the Army's jumbo 100-pound chemical bomb, the so-called M47A2, designed for liquid mustard agent. The bomb was currently in production by the Chemical Warfare Service, and it was known to fragment its 70-pound mustard payload into droplets that were approximately 1,000 microns (1 millimeter) in diameter. That was far larger than what was then thought to be the optimum size for a biological agent (10 to 100 microns, the size of fog particles), but it was something.

However, there were as yet no biological agent production plants operating at Camp Detrick (other than "Black Maria," which was turning out X for the British), and so the munitions men had to make do with an alternative. They chose plain baker's yeast—Fleischmann's, in fact—the kind used in bread making all over the country. Its two virtues were that it was a living organism and it was available cheaply in huge quantities.

So the munitions men put together a makeshift laboratory at one end of the officers' quarters. There they mixed one part Fleischmann's yeast with two parts tap water in a Waring blender and poured it into one of the empty 100-pound bombs until it was full. The bomb took a lot of agent—it swallowed about twenty-two liters of the stuff. Then they inserted the axial burster, connected it to the fuse, and topped it all off with an electrical blasting cap. They took the finished device out to the flat fields between the base and the mountains and wired it up to a blasting machine.

There remained the wee matter of sampling. You did not just stand back and watch while the bomb blew up. They wanted to make a science out of this, the objective being to get a quantitative readout of the dispersed organisms at any given point in the aerosol cloud that resulted. That meant a layout of sampling apparatus, of which, at that point, the Army had none. So the munitions men invented their own.

The first models were evacuated one-liter glass bottles containing a couple of ounces of sterile water. The idea was to open these bottles at the precise moment the cloud passed over them. The vacuum inside would draw in a bit of the cloud, including some of the suspended live organism (assuming any survived the blast), which would end up being deposited in the water at the bottom. By culturing the water's bacterial contents you could get an approximate count of the concentration of suspended organisms per unit of air at the point where the observation was taken.

The trick would be to open the vacuum bottles at the desired instant by remote control, a feat easier said than done. But here the Detrick munitions crew had an inspiration. The vacuum bottles had a sealed glass neck at the top, and the neck was thin enough to be broken by a sharp impact. By positioning the neck under the wire loop of a cocked mousetrap, and by electrically springing the trap at exactly the right moment, the wire loop would snap off the glass neck, the bottle's internal vacuum would draw in a portion of the aerosol cloud within a second or two, and lo and behold you'd have your sample.

Accordingly, rows of five-foot-high wooden stakes appeared on the Detrick cow pasture at distances of 50, 100, and 150 yards from the blast point. At the top of each stake was an open-faced wooden box that contained an evacuated glass bottle, a standard-issue Victor mousetrap, and an electrical mechanism that sprang the trap on command.

And in this way, with the aid of Fleischmann's yeast and 128 Victor mousetraps, Camp Detrick's first open-air biological bomb test took place on July 13, 1943. The bomb erupted and blew a fine mess of yeast slurry into the air and over the test grid, and the mousetraps flew closed and snapped off the bottle tops shortly thereafter.

The test was successful in the sense that some of the samplers had actually pulled in a bit of the aerosolized yeast, but the particle counts proved disappointing. Still, it was a beginning.

Soon they tested other bombs, bigger and smaller, and eventually got rid of the mousetraps in favor of new and improved sampling devices. And once the new pilot plant got going, they were even able to make use of *Bacillus globigii*, which, because it was a spore former like the anthrax microbe, could be used as an anthrax simulant. After several months of testing, however, the Detrick scientists had not found a truly workable biological munition. All the bombs seemed to kill almost every last bit of pathogen.

The British, however, came to the rescue when Lord Stamp arrived from Porton with a copy of the British four-pound Type F biological bomb, plus blueprints.

Trevor Charles Stamp, Lord Stamp of Shortlands, was a British aristocrat whose father, mother, and brother had all been killed in the blitz in April 1941. He'd been trained as a biologist and worked in the Public Health Laboratory, but he felt useless there during wartime. He'd heard the rumors that the Germans were working to develop biological weapons, and he wanted to help with the British effort to retaliate in kind. "I was determined to pay back the Germans for what they did," he explained later, "and to see that our country was not left defenseless as London was when my family was killed."

So he met with Paul Fildes at Porton Down, became a member of the Porton biological division, and took a room at the Old George Hotel along with the others. For a while he experimented with techniques of preserving bacteria by drying—dried pathogens had a longer useful life than those in liquid suspension—but he regarded this work as unexciting. In 1943 Fildes made Stamp the official British liaison to both Canada and the United States, and in the fall of that year he arrived in Washington with blueprints and specifications for the British four-pound biological bomb, a few working prototypes, and a report of the recent trials of the munition on Gruinard Island.

Stamp made some impressive claims for the bomb, in particular that a cluster of 106 of them would put up an anthrax cloud that would be fatal to fifty percent of all humans within a distance of one mile. Subsequent tests by the Americans showed this to be an unrealistic estimate, but nevertheless the device appeared to be highly effective at setting up an aerosol.

So, having been pushed into the germ warfare business by the British from the outset, having adapted their milk-churn bacterial production system, having then filled their order for "three kilo dried X," and lastly having learned how to do open-air trials from them at Gruinard Island, the American germ warriors now adopted the British four-pound bomb as their first and foremost biological munition.

A few improvements would have to be made, however. For one thing, the bomb's interior surfaces had not been adequately protected from corrosion, making the device susceptible to leaks. Accordingly, the Detrick munitions men searched for a corrosion inhibitor that would not inactivate the microorganism, that would adhere well to the iron walls of the bomb tube, and that could be applied without any surface discontinuity—no gaps, cracks, or pinholes. They tried seven different types of coating before deciding on a glassy, baked-on finish made by the Stoner-Mudge Co., chemical manufacturers, of Pittsburgh.

Then there was the problem of end-closure. The British bomb did not have leakproof fittings at either extremity—there was a crimped metal push-in cap at one end and a friction-fit cup at the other—and either one of them or both might let some agent trickle out, with fatal consequences to anyone in the vicinity. The British themselves recommended the liberal application of a sealant cement at the time of assembly—a witches' brew of gum dammar, asphaltic bitumen, coal tar naphtha, and so on—but the

Americans had little confidence in the mixture. In the end Don Falconer, a Detrick physicist, designed a new closure that relied on brazing both ends into place, among other things, and these changes eventually produced an airtight fit.

Still, there had to be a way of checking the security of filled bombs in the field, and someone came up with the idea of adding fluorescein, a fluorescent chemical, to the biological agent suspension. If you shined an ultraviolet light on the bombs, any leaks would be visually apparent. So the munitions men added fluorescein to the liquid at the rate of 1 part per 50,000, a concentration that did not adversely affect the agent, and with this innovation they imagined that they were done.

But then someone realized that fluorescence effects on the ground near an exploded bomb would be a definite tip-off to the enemy that biological bombs were being used against them. This could enable the enemy to undertake medical countermeasures with its troops, defeating the purpose of the biological bomb drop. You would therefore have to dilute the fluorescein to the point that there was just enough of it in the mixture to reveal a leaking bomb, but not enough to show up on the ground after explosion.

The munitions men made this adjustment, too, at which point they decided they had a halfway decent biological bomb on their hands.

But no hot agents to fill it with.

For all their superiority in research and testing, the one area the British had no experience in was truly large-scale pathogen production. The thirteen-gallon milk churns, fine as they'd been as prototypes, were semi-scale vessels at best. When it came to turning out vast masses of industrial commodities, the Americans, inventors of the assembly line, were the experts.

Pathogenic bacteria, however, were not normal industrial goods and had never before been produced in bulk quantities, which meant that the Camp Detrick researchers would have to invent and prove out the necessary hardware, systems, and operating procedures. Creating this maze-works was the special preserve of the Pilot Plant Branch; its members were to create the prototype bacteria factories, which, when perfected, would then be built elsewhere to produce the needed tons of pathogen. Within the first six months of the branch's existence, two pilot plants rose at Camp Detrick, both of them inside the old aircraft hangar.

The location was largely for disguise purposes. Every aspect of the work at Detrick was top secret, but the mass production of hot agents was the most secret of all, and the hangar provided an outer layer of camouflage. It was a light and airy structure with huge windows and whose 50,605 square feet of floor space was large enough to accommodate a couple of four-engine bombers simultaneously. In late 1943, a group of masons invaded the hangar and bricked up much of the interior space with concrete building blocks, creating a hidden inner fortress. The pilot plants would go inside it.

Pilot Plant 1 was long and narrow, about eighty feet by twenty, consisting of ground floor, second floor, and a long steel balcony that overhung the upper level. All of it was completely enclosed by yellow glazed-tile brick walls that had no openings other than for ventilation louvers along one wall and doors to admit personnel and equipment. The bacteria production system was essentially that of "Black Maria," except that the new tanks were more specialized in function, and were increased in capacity and number. There were now different vats for sterilizing the slurry and for cultivating the agent, plus a separate line of fifteen-gallon catalyst tanks for growing the starter culture. Also, the workers now looked the part, decked out in protective suits, black rubber gloves, and hard hats with face shields.

The general flow of materials was much as in "Black Maria," beginning with the workers seeding the six catalyst tanks with flasks of the organism to be produced. Once the seed culture had grown for a few generations inside the catalyst tanks, the workers pumped the contents into fermenters, each of which held 230 gallons of mother liquor. Then, at the end of the fermenter tank growth cycle, they transferred the slurry into two settling tanks, both of which had been salvaged from "Black Maria." After settling, the final product was collected into five-gallon glass carboys as before. Hooded lights overhead illuminated the walkways, feeder lines, steam lines, cooling lines, and assorted other vats, valves, meters, motors, agitators, pumps, sight glasses, vents, pressure gauges, thermometers, air bubblers, and other items, and when fully operational the interior of Pilot Plant 1 looked, smelled, and sounded like nothing else under the sun.

Elaborate as it all was, it was but an overture to Pilot Plant 2, the birthplace of true mass production of hot agent in the United States. Pilot Plant 2 held two 3,700-gallon fermenters and a single 10,000-gallon fermenter. That was exactly the size that Ira Baldwin had spoken of a year previously when he'd said, "I don't know how many 10,000-gallon tanks it will take to produce tons of pathogen, but if you get enough tanks I'm sure you will get tons."

The 10,000-gallon tank was a stainless-steel welded cylinder that extended vertically through all three floors of Pilot Plant 2. It worked on precisely the same principle as all the smaller tanks that preceded it, except that the product did not trickle down into a five-gallon glass bottle. Its output was so large that the separation of spores from growth medium and their final concentration took place in an adjacent facility, Building T-63, the Separation and Concentration Building. Inside it were three 5,000-gallon tanks and three 1,000-gallon tanks together with the final necessary element of the production line, the bomb-filling room. Empty four-pound bombs would come in on pallets, and they'd be shipped out filled with the anthrax agent.

Manufacturing tons of anthrax agent, however, necessitated some special measures in the production process. For one thing, large amounts of "process air" would have to be pumped through the fermenters in order to get the *Bacillus anthracis* inside them to sporulate. That air had to be clean—in fact it had to be sterile—otherwise you'd be growing some stray bacteria along with, or even instead of, the target microbe. Ordinary unfiltered air, the stuff you inhaled every moment of every day, was chock full of microorganisms: pollen grains, seeds, tiny bugs, parasites, floating funguses, flying viruses, and countless bacteria of every description. All that would have to be eliminated; the very air itself, in other words, would have to be sterilized before use.

Air sterilization, however, was a new technology, particularly in the high volumes needed for the desired tonnage of anthrax spore growth. To solve the problem, the Detrick engineers first ran the incoming air through stacks of glass-fiber filter mats to remove the dust particles and then forced the air through a sophisticated electrostatic filtration system to eliminate everything else. Such treatment left the air ninety-nine percent free of all microorganisms. That was still not good enough for mass-production purposes, however. Microorganisms replicated like rabbits and small numbers of the wrong sorts of germs could ruin an entire production run—as, indeed, would often happen.

The engineers therefore fell back on heat sterilization, warming the incoming air to temperatures between 380°F and 400°F. Heating the process air killed the microorganisms, but it also made for an extra step in the form of cooling, for the sterilized air had to be chilled to about 90°F before it was pumped through the fermenter tanks under pressure. To cool it off the scientists brought in cold water from nearby springs and ponds and sprayed it over the exterior surfaces of the air ducts.

All that was for *incoming* air and slurry; a separate set of procedures was required for the *outgoing* air and slurry, both of which, in the course of use, had become contaminated with the anthrax bacillus. The process air was burned—literally incinerated—by an "air incinerator," an industrial oil burner. The contaminated air was forced into the combustion chamber where any microbes present were burnt to a crisp. The air that finally emerged from the Detrick smokestacks was far cleaner than the air that went in.

The contaminated growth medium, for its part, was transferred to a slurry decontamination station where it was steam-sterilized, not once but twice, after which it was chemically tested and finally let out into the city sewer system.

That left the anthrax spores themselves, the production of which had been the object of the entire exercise. The slurry carrying them was sent through stainless steel pipes into Building T-63, adjacent to the hangar, where the spores were separated from the mother liquor and then concentrated into a deadly essence.

Before loading it into bombs, however, the scientists first had to determine the agent's virulence.

From an offensive germ warfare standpoint, the major practical question concerning a given batch of hot agent was its ability to cause disease. Other things being equal, the more virulent the agent, the better, but of course there was no way to test virulence on humans, not in America, at least not yet. Camp Detrick, furthermore, was near a small town, which meant that there was no way to test hot agents in open-air field trials, even on animals. The scientists therefore proved their products in "cloud chambers," which was to say, Henderson tubes and their direct descendants. So along with barracks buildings, test grids, and pilot plants, the Army built a succession of laboratory buildings to house the pathogen research facilities at Camp Detrick.

Each lab building was a duplicate of the next. From the outside they were long, low structures made out of yellow glazed building blocks and resembled the dormitories of a juvenile offender detention center. The buildings, which were about fifty feet by a hundred, contained offices, clothes-changing rooms, cloud chamber rooms, dissection suites, and holding cells for the animals. The interior was divided into "clean" areas and "hot" areas, with offices and changing rooms on the clean side and everything else on the hot side, the two areas being interconnected by an airlock. The entire complex was vented by blowers that drew in outside air, maintained negative pressure throughout the building, and exhausted the contaminated gases to the air incinerator located in an annex at the back.

To minimize the chance of cross-contamination among the various types of bacteria to be studied, the Army built several identical laboratory buildings at Camp Detrick and assigned each one to a different agent. One lab building was devoted to "N" (*Bacillus anthracis*), another to "US" (*Brucella suis*, which caused brucellosis), another to "UL" (*Pasteurella tularensis*, the cause of tularemia), and so on down the list. Eventually there were about a dozen separate dedicated lab buildings in the restricted area at Detrick.

The numbers of animals required for research work on all those different pathogens prompted some innovations in the matter of cloud chambers. Henderson tubes, in the view of the Detrick experimenters, left a lot to be desired. In the Henderson apparatus only the animal's head was exposed to the pathogen cloud, which was not at all true to life: in a real-world biological warfare situation, the subject's entire body would be exposed to the agent. Second, an animal immobilized in a Henderson tube was scared out of its wits, puffing and panting and ventilating itself like crazy, which also gave rise to misleading results. Worst of all, the British units were capable of exposing only a paltry two or three specimens at a time, placing an intolerable drag on the further progress of aerobiology.

The Americans therefore developed a new and improved cloud chamber in the form of a modified autoclave. The first units were about the size of a microwave oven, but they could easily hold as many as a dozen mice at a time, and all of them could be exposed to the agent simultaneously. Since it was already an airtight container, the only changes necessary to convert an autoclave into an animal exposure chamber were the addition of a nebulizer to produce the aerosol spray and a circulation system to set up a uniform and continuous flow of the cloud over the animals.

The cloud chambers got bigger and better in the standard American fashion until acceptable numbers of animals—twelve guinea pigs, sixteen rats, or thirty-six mice—could be exposed at a time. Later units, known as Reyniers chambers after their inventor, J. A. Reyniers, of Notre Dame

University, were equipped with a Pyrex sight glass that allowed you to watch or photograph the test subjects as the agent cloud passed over them and with glove ports so that you could handle the animals during the course of a test.

A lab animal about to undergo what was politely called an "exposure" was in for one of the more bizarre experiences of its short life, and close to the last one it would ever have. The animal, a hamster, for example, would have been raised at the Detrick animal farm, a compound of concrete-block huts on the west edge of the campus. A day or so before the test it would be transported to the lab building, where it would live in a metal cage in the animal room. On the day of exposure it, along with a half dozen or so of its mates, will be put into a "transfer box," an airtight sheet-metal cube with welded joints, sealed doors at both ends, rubber glove ports at the side, and a clear viewing port at the top. From the time it first enters the transfer box until the end of its life, the animal will be in a separate air system from that of the experimenters so that the humans will never breathe in hot agents exhaled by the animals.

The transfer box is equipped with handles at the ends, and one of the experimenters now lifts the box with the animals inside it and carries it to the cloud chamber. This is a stainless steel cylinder two feet in diameter and four feet long with an airlock at one end, a viewing glass centered at the top, and left- and right-hand glove ports along the side. Empty black rubber gloves dangle into the chamber interior like collapsed balloons. Externally, the chamber resembles an iron lung, and there are various pipes, hoses, stopcocks, and other fittings attached at several points along the outer surface.

The experimenter attaches the transfer box to the outer port of the cloud chamber's airlock. Resting inside the airlock is an "animal exposure basket," a wire mesh cylinder large enough to accommodate all the animals together. The inner port of the airlock opens directly into the hollow core of the cloud chamber, and a few minutes before the start of the test, the door between the transfer box and the airlock opens and a human hand in a black rubber glove guides the animals into the wire basket and then closes the door behind them.

It is absolutely dark inside the airlock.

For a period of ten minutes, during which time they're expected to settle down, collect themselves, and start breathing normally, the hamsters are ventilated with clean air. Inside the cloud chamber the test pathogen has already been aerosolized.

At time zero, the airlock's inner door opens and a hand inserted through a glove port pulls the animal basket out of the airlock and into the chamber itself.

For the next ten minutes, the hamsters inhale the aerosolized hot agent. If one of them happened to look up, it would see a human face covered by a white gauze surgical mask observing them through the sight glass and shining a flashlight here and there around the interior.

The animals come out through the reverse procedure.

The six hamsters are now "hot"; all of them are exhaling trace quantities of the pathogen, and so they will spend the rest of their days in ventilated cabinets, hermetically sealed metal enclosures, each with its own air, water, and food supplies.

Every morning, a technician lifts out the dead bodies and brings them to a safety cabinet for autopsy. He cuts each animal open, removes its liver and spleen, and drops the organs one by one onto a petri dish.

The carcasses then go to the autoclave to be steam-sterilized. Then they go to the incinerator. There they go up in smoke.

5

The United States clearly needed its own Gruinard Island. Cloud chamber tests, helpful as they were in assessing the virulence of biological agents, were still no more than lab experiments performed in highly artificial and controlled environments, and they gave little or no direct evidence as to the military effect of those same agents when dispersed by a biological munition on the battlefield. For that, there was no substitute for open-air trials with biological bombs.

In November 1942, after they'd heard about the first wave of Gruinard trials from Fildes and Henderson, officials of the Chemical Warfare Service had started casting around for a likely site for field tests. For obvious reasons a remote island seemed like an ideal location, so they considered potential sites on Pooles Island, Barren Island, and Fisherman's Island, all in Chesapeake Bay; the Sea Islands off the Georgia coast; Mullet Key and Egmont Key in the mouth of Tampa Bay; Dry Tortugas, a former pirate hideout west of the Florida Keys; and Ship, Petit Bois, and Horn Islands in the Gulf of Mexico off the coast of Mississippi.

On January 19, 1943, Ira Baldwin, together with two Chemical Warfare Service officers, left Washington on a train for New Orleans to tour the islands in the Mississippi Sound. They made an aerial survey of five islands and later visited three of them by boat.

Horn Island was the biggest of the three, a thin spit of land about ten miles long and barely a mile wide at its broadest point. It was isolated and self-contained, accessible only by private boat across a seven-mile stretch of the Gulf. Civilization, such as it was, lay in the cities of Biloxi and Pascagoula, about fifteen miles across the water.

The island itself consisted of white beaches, low scrub, and sand dunes covered with long grasses and sea oats. As Baldwin and the others walked along the shore, they saw bales of rubber here and there on the sand and black oil slicks floating on the waves. These were fleeting traces of the war in Europe; German submarines cruising the Gulf of Mexico had torpedoed American cargo ships, and this was part of the wreckage.

Inland there were loblolly pines and tall spindly oaks, plus a network of small brackish ponds and lagoons. A family named Waters had lived and raised cattle there in the distant past, and a few tracts on the island were still privately owned, but there were no longer any human habitations that they could see, and other than for a herd of wild ponies the place was a deserted semitropical island. By the end of March the Army had acquired all of Horn Island for the exclusive use of the Chemical Warfare Service.

The island ran in an east-west direction, and the Army put up a north-south fence across it at about midpoint, confining the wild horses to the east end until they could get them off the island. West of the fence they constructed a set of corrals for the test animals, including separate enclosures for pigs, goats, and horses, plus rabbit hutches and assorted holding pens for other species.

The grid area, where the biological bombs would be detonated over lines of animal subjects, was five miles further west. So many animals would have to be transported from the corrals to the test grids, and the trips would have to be made so frequently, that the Army decided to build a railroad line between the two points. Accordingly, a detachment of Navy Seabees arrived and laid 7.6 miles of narrow-gauge railway track along the island's north shore. Later, they shipped in two 14-ton steam locomotives and twenty 10-ton wooden freight cars from Fort Benning, Georgia, and took them for trial runs back and forth across the island.

And all that was just for the animals. For the 200 or so test personnel to be stationed there, the Army built a set of barracks and recreation buildings for the officers, a second set of the same for enlisted men, a headquarters building, a laboratory building, a mess hall, a concrete basketball court, a machine shop, and a small theater. They finished it off with the necessary utilities, including a power plant, car and truck garages, storage sheds, water and sewage systems, and an incinerator. Within a year of Ira Baldwin's having first set foot on Horn Island, the Army Corps of Engineers had converted this deserted island paradise into one of the world's major biological warfare test centers, with a total of 144 separate structures, its own railroad line, and a fifty-foot-high pole flying the American flag.

The test procedures were equally elaborate, and the Army had prepared detailed, minute-by-minute timetables that coordinated the various crews and their functions and listed the scheduled times of train arrivals and departures at the successive stops, with all the events converging at time zero, when the bomb exploded.

Test day began with the animals being loaded aboard the Horn Island railroad line for their one and only ride to the grid area. With the animals aboard, the train made a stop at the administration area to pick up the day's test range personnel, then made a second stop at the dressing station, where the men climbed into their double-layer protective suits, gloves, boots, and gas masks. The locomotive then started up again for an uninterrupted ride along the shore to the test grid. The five-mile trip along the north beach was pleasant and bracing, and on a good day the men could see the Pascagoula skyline in the distance.

The grid area was a flat sandy expanse of land with little vegetation on it other than thin grasses and a few shrubs, and with scant trees to interfere with wind currents. Here, as on Gruinard, the animals would be set out in long lines and then, at time zero, a British four-pound bomb filled with hot agent would be detonated upwind, after which the cloud of pathogen would wash over the animals and then blow out to sea.

Within minutes, the animals would be taken to the operations area, a bunch of corrals and lab buildings on an elbow of land that jutted out slightly into the Mississippi Sound. There they would wait out the incubation period of the test organism.

Autopsies would take place in the lab building, a wooden structure with a long center hallway inside and a steel monorail track that ran the length of the ceiling. It was the kind of overhead track that you might find in a slaughterhouse or meat-packing plant. The procedure was to hang the carcasses on hooks and then push them along the track in assembly-line fashion while technicians working in protective gear cut open the hides and took the necessary blood and tissue samples. The carcasses would then go to the incinerator to be burned and the ashes buried in a disposal pit.

That, at least, was the plan.

After its first full year of operation, Camp Detrick was beginning to look like a semipermanent military installation. The temporary wood-plank side-walks, unstable slats that squished under your feet, had been replaced by conventional concrete walkways. Buildings had gone up in a headlong rush, and by the spring of 1944 there were roughly a hundred structures on the post in various stages of completion. The Army had built, equipped, and successfully operated a dozen or so bacterial laboratories, two major pilot plants, and an animal farm, giving the antiquated airfield the look and feel of a latter-day biological research center. Over the months Ira Baldwin and his colleagues had put together a force of about 1,000 officers, enlisted men, and civilian researchers, and had collectively charged them with the task of turning biological warfare into a rational and orderly scientific discipline.

In this they had made considerable progress. The scientists had demonstrated that X, botulinum toxin, and N, the anthrax bacillus, could be cultivated in any arbitrary volume, whether by the ounce, pound, or ton, so long as you had fermenter tanks of the appropriate size. They had developed techniques and instruments for sampling bacterial clouds. They had detonated a few biological bombs and gotten some modestly encouraging results. They'd tested a variety of airborne pathogens on hundreds of animals of various species, and had developed some notions as to what constituted the so-called LD_{50} for each, the median lethal dose, the amount that was fatal to fifty percent of the animals exposed to the agent. They had proved that air incineration was a practical and highly efficient decontamination procedure.

Above all, they had done this safely, with no human deaths and very few accidental infections. There had been some: Newell Johnson, a physician who had worked in the pilot plants from day one, had one morning noticed a rough spot on his elbow, a dim reddish-brown patch about the size of a quarter. The pilot plant had been growing one of the first batches of anthrax organism, and the rash was almost certainly the sign of cutaneous anthrax. But it had responded well to penicillin, and Johnson recovered without any problem.

Another time, one of the reactor tanks in Pilot Plant 2 started overflowing with anthrax slurry. The stuff had entered an air vent and was pouring out of the building like suds from a washing machine. Three workers from the E (Engineering) Division grabbed shovels, flung up a dirt barrier around the outflow, and stopped it before it ran into a storm sewer that emptied into a creek that eventually wound its way through the cow pastures.

Detrick officials, however, were well prepared for any catastrophe. There was a forty-eight-bed hospital on the post, and there were plans for an autopsy room, a pathology room, and a rank of body coolers. And, of course, there were contingency procedures for the disposition of infected remains. In March 1944, General William N. Porter, chief of the Chemical Warfare Service, had asked the Judge Advocate General of the Army to say whether in the case of a death caused by an infectious agent, the decedent's infected remains could be disposed of secretly. The answer was yes: "Deceased personnel might be placed in a hermetically sealed metal casket and interred by military personnel in the area, without disclosing by certificate, report or statement the nature or cause of death."

Secrecy was always an uppermost consideration at Detrick, and people working in one lab building were prohibited from talking about their jobs with those in the next lab, or, indeed, with anyone else. To ensure compliance, the Army stationed spies in the barracks to eavesdrop on conversations and report infractions. Talking shop outside the post, and especially on the streets of Frederick, was forbidden, and the Army monitored adherence to this, too, sending out intelligence agents for periodic "security surveys." In May 1944, one agent reported back that "little interest was taken by the townspeople of Frederick in the activities carried out at Camp Detrick. It was general knowledge among townspeople that Camp Detrick is a secret chemical warfare installation and it is believed by them generally to be a research installation where a new secret gas or gases are being developed. . . . However, it is this agent's opinion that anybody who really wanted to find out that bacteriological warfare activities are being conducted at Camp Detrick could easily do so by studying the background of the technician civilian employees employed there and by analyzing the type of material purchased by the post procurement in Frederick and the types of material which is shipped into Frederick by Railway Express."

On the post, meanwhile, the Army had set up a "Special Projects School," its mission being to give students "an understanding of the known technical facts and potentialities of germ warfare." The place had all the trappings of a genuine educational institution, with faculty, course offerings, credit hours, and classes that included lectures, demonstrations, and "field trips" to laboratories and pilot plants. The first graduating class of sixty students took a three-week course of seventy-three credit hours that covered:

Significance of BW and possible enemy tactics	3 hours
Intelligence	3
Microbiology	5
Immunology	6
Agents	9
Production of agents	4
Munitions	12
Food and water contamination	3
Methods of detection	2
Clinical and laboratory diagnosis	3
Physical protection	9
Epidemiological control	14

A school spirit developed quickly enough, and by the start of the third "semester" there was a class motto, colors, and yell. The class motto was: "We seek something which cannot be seen, smelt or felt, discovered by means which we do not have, and to be cured by something which we make from nothing, not later than yesterday." The class colors were globigii yellow and methylene blue. The class yell was:

> *Brucellosis, Psittacosis*
> *Pee! You! Bah!*
> *Antibodies, Antitoxin*
> *Rah! Rah! Rah!*

And, as befitted an emerging hotbed of theoretical and applied science, there were internecine scholarly disputes on key issues. Some of them were highly emotional, and an early discussion of psittacosis microbe as a possible germ warfare agent nearly led to a brawl among the participants.

On the one side were D. W. Watson and L. C. Kingland, both of whom claimed that psittacosis (parrot fever) had greater offensive possibilities than either botulinum or anthrax. A natural outbreak of psittacosis in Louisiana, they claimed, had caused an epidemic that was fatal to eight persons out of nineteen cases. Plus, the disease had passed through six consecutive person-to-person transfers without any loss of virulence, suggesting that the agent would be capable of causing a genuine person-to-person mass epidemic that would kill or incapacitate a lot of people.

Those in the audience, including Lord Stamp, who was there visiting from Porton Down, explained that no work on psittacosis had been authorized, and that Camp Detrick therefore had to continue to concentrate on anthrax. A shouting match ensued, and the incident became the subject of a formal "Report of Occurrences at Technical Meeting and of Subsequent Disciplinary Action." Both Watson and Kingland were reprimanded for their outbursts.

Evidently, it was not easy working with hot agents on a daily basis, and a number of Camp Detrick technical personnel ended up as "psychoneurotic cases" and were dismissed. "The nature of the work on the post," said the Army, "was not considered conducive to rehabilitation."

So far, biological warfare was still essentially just a concept, a realistic possibility but not a clear and present danger. No nation had ever used it in battle, or had threatened to, and so for all the research that everyone concerned had put into it, an actual germ warfare attack remained a distant and somewhat academic prospect.

All that changed in December 1943, when the news arrived in London and Washington that the Germans were readying a biological weapon for use against the Allies, a pilotless plane or rocket called the V-1. According to U.S. Army intelligence reports, the V-1 rocket would carry a warhead filled with botulinum, the most toxic substance known to science, which, if intelligence reports were correct, would soon be falling in great quantities upon the streets and pastures of England. In that event, the Allies would have to defend their troops against the substance and then respond in kind, retaliating with a biological bomb of their own.

Defense meant inventing and mass-producing a toxoid that immunized the human body against botulinum. In 1943 at Harvard, Alvin Pappenheimer and Howard Mueller had already created a workable toxoid against botulinum and had tested it out on mice, guinea pigs, and, finally, themselves. Later, Pappenheimer and his crew at Camp Detrick worked out a technique for mass-producing the substance, and by the summer of 1944 members of Detrick's D (Defensive) Division had manufactured and stockpiled more than 4,000 gallons of botulinum toxoid, enough to immunize approximately 700,000 troops.

Offensive work was not quite so advanced, but in the spring of 1944, the British placed a formal order with the American government for biological bombs. The order had come from Prime Minister Winston Churchill himself, who had instructed Ernest Brown, chairman of the British Bacteriological Warfare Committee, to request 500,000 anthrax bombs from the Americans. "Pray let me know when they will be available," Churchill wrote in a memorandum. "We should regard it as a first instalment."

But although American military contractors could mass-produce that quantity of empty bombs easily enough, filling them all with anthrax spores was another matter. For all its proof-of-concept work with large fermenters, Camp Detrick had not yet gone beyond the pilot plant stage when it came to producing hot agents in the quantities needed to fill half a million biological bombs. Moreover, Detrick's plants were *pilot* plants, not *production* plants, meaning that they were only test beds or prototypes rather than true mass-production facilities. The British themselves, who had never progressed beyond their simple milk-churn production system, were likewise in no position to manufacture large volumes of pathogens. That left Canada.

In the summer of 1942, while Detrick Field was still an abandoned Army Air Corps flying strip, the Canadians had cobbled together a bacterial production system of their own on Grosse Ile, an island two miles long and a mile wide in the St. Lawrence seaway east of Quebec. Earlier, Grosse Ile had been an immigrant quarantine station, the Ellis Island of Canada, where passengers who arrived with communicable diseases waited out the course of their illnesses, but it had been abandoned for many months. The island was in midriver and totally unconnected to the shore, and, being uninhabited, had all the basic prerequisites of a germ warfare research installation.

On July 21, 1942, a group of Canadian bacteriological warfare researchers accompanied by three Americans made a reconnaissance tour of Grosse Ile. From their point of view, the focal point of the abandoned complex was the disinfection center, a building in which the immigrants had been deloused and, in effect, steam cleaned. It consisted of two large rooms separated by a walk-in autoclave. The incoming passengers removed their clothes in the outer room, piled them in the steam chamber, and then walked through a bank of hot showers. By the time they and their clothing emerged on the other side, all were spotless.

The steam chamber could be sealed off so as to create an airtight enclosure for pathogen production, and the entire setup seemed tailor-made for germ warfare purposes. The Canadians decided to convert the chamber to the production of anthrax spores, which they hoped to grow at the rate of 300 pounds per week.

The American government agreed to provide seventy-five percent of the Grosse Ile operating expenses, and by the summer of 1944 the place was churning out a modest supply of the N bacillus. Still, the Canadians lacked the American gift for mass production, and so instead of using large-scale batch methods such as the 10,000-gallon fermenter tank at Camp Detrick, the Canadian scientists were making essentially laboratory quantities of the agent, although in multiple amounts.

They grew the spores in small bacterial culture trays of the size that might be found in a college biology lab. Technicians poured growth medium into the trays by hand and then seeded each one with starter culture. At the end of the growth period they drew off the new spores with a small metal rake, much as a chef skims grease from a pan.

The Canadian scientists installed hundreds of such trays in the steam chamber and at one point they had 1,280 of them producing bacteria simultaneously. Still, the procedure was unsafe for the workers, the total output was small, and within a few months of having started, the Canadians terminated anthrax spore production at Grosse Ile.

The Americans, meanwhile, had already gone ahead with a separate bacterial factory of their own. On June 20, 1944, in the wake of the British order for 500,000 filled anthrax bombs, the Special Projects Division of the Chemical Warfare Service issued SPD Manufacturing Order No. SP-1. The order called for a total of one million Mk I bombs, the American version of the British four-pound Type F bomb, to be mass-produced, filled with anthrax, and packed into clusters. Half would go to the British; the other half would be retained by the United States for possible use.

The bomb casings and other internal components would be manufactured by the Electromaster Corporation, a commercial bomb maker in Detroit, Michigan. The high explosives—the pentolite pellets and tetryl powder—would be made by the Unexcelled Manufacturing Company of Cranbury, New Jersey. The anthrax spores, the biological core of the bomb, would be manufactured by the U.S. Army at its new production center in Vigo, Indiana.

Vigo was a small town six miles south of Terre Haute, surrounded by cornfields, hog farms, and flat plains. In 1942, prior to getting into the biological warfare business, the Army Ordnance Department had constructed, on a 700-acre tract near Vigo, a $21 million factory complex for the production of conventional munitions—the standard high-explosive bombs that were being dropped out of aircraft by the million. But it was soon clear that the Army had greatly overestimated its munitions needs, and so in July 1943, scarcely a year after it had been built, the Army decommissioned the Vigo Ordnance Plant and leased out portions of it to the Delco Radio Corporation for the manufacture of military electronics equipment.

That arrangement lasted for all of ten months—until May 1944, when the Army's Special Projects Division, which controlled the biological warfare project, decided that it needed the space and facilities to fill the British anthrax bomb order. Shortly after an inspection team toured the Vigo plant, the H. K. Ferguson Construction Company came in and added the necessary fermenter tanks, air compressors, refrigerators, and slurry heaters. They converted buildings to biological laboratories and built catalyst and separation buildings, a sewage decontamination plant, and an animal farm.

At the heart of the complex were the anthrax fermenters, which, at 20,000 gallons each, were larger than even Ira Baldwin had ever contemplated. The tanks were about twenty feet wide and forty feet high, and twelve of them were installed at Vigo. The total production capacity at the Vigo plant was therefore 240,000 gallons, making it the largest bacterial mass-production line ever created anywhere in the world.

Getting all that anthrax slurry into one million individual bombs, however, would prove to be a major bottleneck. Obviously, some sort of mechanical filling apparatus was required, and so members of the M (Munitions) Division at Camp Detrick made a survey of commercial high-speed filling machines commonly used for loading bags, cans, and bottles with flour, starch, baking powder, beer, milk, and the like. Supposedly, all of these machines were "dustless" or "dripless." When they got out into the real world and saw such devices in action, however, the Detrick munitions men found that a machine was considered "dustless" if it lost no more than a pound or two of filling a day. Bottle-filling machines, likewise, were called "dripless" so long as no more than a quart or so of liquid fell through the cracks. Even the machines that the pharmaceutical industry used for filling bottles with drugs or vaccines gave off small amounts of invisible spray. None of this was tolerable for a biological bomb operation.

The British, on the other hand, had developed a bomb-filling machine to load hot agent into its own four-pound Type F bomb. Since the American Mk I was almost identical to the Type F, the British sent a prototype of its bomb-filling machine to Camp Detrick, where the munitions division men put it through tests.

The device worked somewhat like a drill press. The operator placed an empty bomb casing on a platform and then pressed a foot pedal that lowered a filling head onto the open end of the bomb. The filling head created a slight vacuum in the bomb chamber and also fed slurry into the bomb through a nozzle. Then, when the bomb chamber was full, the operator released the foot pedal, the filling head rose up and away, and the operator lifted the filled bomb off the platform.

The advantage of the British machine was that it was fast: a bomb could be filled with agent in fifteen seconds, making for a production rate of 240 bombs per hour. The disadvantage was that the procedure was not spray proof: the slurry liquid foamed up as it poured into the bomb, bubbles popped open, and spray particles went all over the place to a radius of twenty feet. Even with the operator dressed in protective clothing, gloves, and gas mask, this was unacceptable.

The Detrick crew now considered three possible solutions to the problem. One, they could make the agent nonfoaming. Two, they could pour the liquid in without splashing. Three, they could place a rubber diaphragm over the open end of the bomb, puncture the diaphragm with a hollow needle, send the slurry in while simultaneously removing the air displaced by the slurry, and then withdraw the needle. This was clearly the most promising, but it would take engineering, development, and test work, plus large quantities of rubber, a commodity in short supply at the height of the war.

And so to get on with immediate mass production—the war was now three years old and still not a single filled biological bomb had rolled off an American assembly line—the officers of Detrick's M Division decided to use the British filling machine just as it was, except for the addition of a hood and flue arrangement that would draw off the surrounding air and take the pathogen spray particles with it. Besides, the operator could be placed in an airtight suit. The Camp Detrick engineers built six of the British machines and shipped them off to Vigo.

The production line after that was straightforward. The filled bomb casings would go to the bomb assembly building, where each would be fitted with a detonator, after which they'd be taken to the separate cluster assembly building. The four-pound bombs were not meant to be dropped out of a plane one by one; rather, 108 of them were to be ganged together like a bunch of asparagus and fitted into an M26 cluster adapter, essentially a large empty can with tail fins at the end. The final product, the 108 filled Mk I bombs loaded inside the cluster adapter, was formally known as the M33 biological cluster bomb, total weight 500 pounds.

The M33 cluster bomb would be what actually dropped out of a plane and fell toward the enemy. An explosive charge at one end would eject the 108 Mk I bomblets at a predetermined height above the ground and scatter them over a large area.

Once Detrick had shipped the six British-designed bomb-filling machines to Vigo, the only thing left to do was to prove the safety and security of the Vigo bacterial mass-production system. Everyone involved wanted the Vigo plant to be safe, so there would have to be shakedown tests. To perform them, Ira Baldwin selected Walter Nevius, a trusted Detrick specialist in pathogen containment techniques. Nevius was cautious and exacting and left nothing to chance, and if anyone could get the production operation going with a minimum of risk, he could.

In the summer of 1944, Walter Nevius rolled into Vigo.

Nevius would follow a gradual, staged, stepwise plan for the mass production of INK-B, the Vigo code name for anthrax. He'd run the system with water first and check for leaks, filling even the Mk I bombs with nothing but plain tap water. He'd find the leaks and seal them off and then run the whole system again, this time using *Bacillus globigii*, the anthrax simulant, as the test agent. Then, at last, if everything was sealed tight as a drum, he'd begin an experimental INK-B run, but even then only at ten percent of plant capacity.

During the course of INK-B production, furthermore, test swabs would be taken at all joints, flanges, valves, and at every other conceivable leak point along the slurry plumbing and the process air shafts. In addition, a modified mine-canary system would be employed as a safety check, with small flocks of sentinel sheep located at strategic points around the restricted area.

The safety tests extended well into the fall of 1944. As they continued, some strange devices from Japan started turning up off the west coast of the United States. The first one, a white blob in the water, was spotted by a Navy patrol boat cruising sixty miles southwest of San Pedro, California. The object consisted of a large mass of wet paper attached to shroud lines at the other end of which was a small metal canister. It was apparently a high-altitude balloon that had traveled clear across the Pacific. The canister's contents had been dissolved by seawater, but it might well have contained a bacterial payload. The Navy shipped the canister to the War Department in Washington, who sent it to the headquarters of the Chemical Warfare Service at Edgewood Arsenal, who sent it on to Camp Detrick for analysis. Should the object have held biological organisms—such as the rice grains, fleas, and plague bacteria that had fallen over China four years earlier—then the United States might be compelled to respond in kind, by dropping bacterial bombs upon Japan.

Well into the spring of 1945, though, the Vigo plant was still deep in safety testing and was by no means ready to start manufacturing pathogen. Not until April 1945 did Walter Nevius pronounce the plant watertight. Not until June did he even let the simulant runs begin.

In July, however, the first atomic bomb was exploded at Trinity Site. In August, the Americans dropped the bomb on Hiroshima. And on September 2, 1945, World War II ended.

6

In the end, none of it was ever used. World War II came to its grand finale not with bacteria spreading over England, Germany, or the United States, but with mushroom clouds hanging over Japan. Not once during the war did any of the main players direct any of its carefully cultivated germs against the enemy.

When the V-1 rocket, Hitler's "secret weapon," finally streaked across the English Channel and hit London, it carried only conventional explosives, no germ warheads. Germany, in fact, hadn't had much of a biological warfare program to begin with—even though it had been British fear of a German biological attack that led them to establish their own germ warfare project at Porton Down, undertake the Gruinard tests, and ask the Americans to mass-produce hot agents and bombs filled with them. All of it had been in defense of a nonexistent threat.

Contrary to Allied intelligence reports, the German biological warfare project was a minor affair that never produced a practical weapon. In the late 1930s, researchers at the German Military Biological Institute in Berlin made a study of anthrax for possible offensive use, and they had also investigated plague, cholera, and typhus, but at no point had they converted them into weapons systems. The reason for this lapse, some historians later claimed, was that Hitler, having been temporarily blinded by British mustard gas in World War I, opposed both chemical and germ warfare and gave orders prohibiting development work.

Japan, by contrast, had embarked upon a massive germ warfare program long before the start of World War II, but both the Americans and the British remained largely ignorant of the Japanese work until the end of the war. The Japanese had produced supplies of lethal agents, had tested them on live subjects, and had developed and tested a variety of biological bombs. And while they had done plague experiments in China, even they had never used any of those weapons on the battlefield.

Which was just as well, since neither the British nor Americans had produced an inventory of filled biological weapons with which to retaliate in kind against Japan, Germany, or anyone else. The United States, in fact, for all of its research, experimentation, and long experience with assembly line manufacturing techniques, had not mass-produced a single filled biological munition after almost two and a half years of crash development work. Moreover, neither Horn Island, which was its principal field test center, nor the Vigo plant, its bacterial production facility, had performed as intended.

Horn Island, in the beginning, had been earmarked for a staff of 200 officers and enlisted men together with a vast panoply of large and small animals. With its packing-house-style overhead tracks for processing animal carcasses, its postmortem facilities were to be the last word in accelerated dissection. The operations area was so large in scale that the Army had constructed seven powder magazines to hold the explosives needed for all the bombs to be detonated on the test grids. Horn Island, in sum, was to be the workhorse site for American biological weapons trials.

But before construction had even started, before the first mile of narrow-gauge railroad track had been laid across the sands, an advance team from the Army Corps of Engineers discovered that large numbers of fishing vessels regularly plied the waters between Pascagoula, Biloxi, and Gulfport and the fishing grounds out in the Gulf of Mexico, and that no warnings, threats, or polite requests from the Coast Guard were successful in stopping them. There had been some high-level discussion, in the wake of this news, of canceling the project entirely and abandoning all plans for using the island as a biological warfare test station. But the necessary funds had already been appropriated, and nobody in government ever turned down money when offered. Work would therefore proceed, but with the understanding that no bacterial tests would be conducted on the island. Open-air test work would be limited to toxins, substances that, because they did not replicate, could not cause epidemics. Soon, though, the Army's meteorologists decided that the area's winds, which blew toward the mainland for two thirds of the year, were not ideal for open-air tests of toxins, either.

The Horn Island Chemical Warfare Service Quarantine Station nevertheless opened for business on October 28, 1943. In its nine months of existence as a biological test center, the Army conducted only twenty-three germ warfare trials there, all of them with botulinum toxin dispersed out of the Mk I.

The trials were not notably successful. A total of fifty-four Mk I four-pound bombs filled with X (botulinum) slurry were fired at the Horn Island test grid over stands of boxed guinea pigs. The bombs were fired singly, and then in combinations of two, three, and four bombs detonated simultaneously, but in no case did any of those bomb combinations kill so much as a single guinea pig by the inhalational route, and in fact no trace of the toxin could be detected in the lungs of the animals during postmortem. The only guinea pigs who died of botulinum poisoning were those that had licked the toxin off their own fur.

In desperation, the scientists then exploded thirteen Mk I bombs simultaneously over a stand of fifteen animals. In that case, finally, one guinea pig died from inhaling botulinum. The Army concluded, in view of these results, that the Mk I four-pound bomb filled with botulinum slurry "would probably not be a lethally efficient weapon."

So ended the Horn Island biological bomb trials. The Army declared the site "excess" on August 13, 1945, and shipped all the lab equipment and unused materials and supplies back to Camp Detrick. The rest of the surplus property—the railroad tracks, buildings, and land worth an estimated $448,000—was turned over to the Corps of Engineers for disposal.

The Vigo plant, at the end of the war, had still not gotten beyond its shakedown safety tests and had manufactured nothing more than *Bacillus globigii*, the anthrax simulant. During the summer of 1945, the Vigo scientists had grown 8,000 pounds of the stuff—four tons—in a single production run. The simulant itself could not be used as a weapon; other than for its role in proving out the production process, it was an entirely worthless microbe, and all of it was therefore destroyed. Still, the Vigo crew had demonstrated that they could grow bacteria "by the ton."

At an earlier stage in the safety testing, General Rollo Ditto, commander of the Special Projects Division, had gotten so exasperated at all the delays at Vigo that he called Ira Baldwin to Washington to see if the hot agent production work couldn't be speeded up somewhat.

Baldwin had been the guiding light of Camp Detrick from the beginning. He'd chosen the site, planned the labs, staffed them, and had a hand in the design and execution of every major operational detail. He was all over the place, everywhere and nowhere, with a finger in every pie. He was especially strict about worker safety, and once had a run-in on the subject with David Henderson, who during one of his periodic visits to Camp Detrick had complained about all the S (Safety) Division's rules and regulations that, in his view, were excessive and hampering progress.

"I've handled these organisms for a long while and I'm quite willing to take any risks that might be involved," he told Baldwin.

"Well, Dave, I'm not really worried about whether you get killed or not," Baldwin said. "If you do, we'll feel sorry about it and we'll take a couple of hours off and we'll go to the funeral and we'll come home and go to work again. But if we get organisms out into the air and Farmer Jones's cows over here get anthrax and they die, we'll have a congressional investigation that will probably shut down the whole post. So I really am not as much interested in you as I am in protecting the community."

After this outburst David Henderson never said anything more to Ira Baldwin about circumventing the safety rules. So when General Ditto told Baldwin that Walter Nevius would have to be relieved of his position at Vigo and be replaced by a chemist, Baldwin was equally unmoved.

A chemist wasn't qualified to run a biological operation, he said. A chemist could tolerate a few leaks in the system—they could just be wiped up, hosed down, and forgotten about. But you couldn't take the same attitude with biological pathogens. Microbes were living entities, they floated through the air, landed on things, they replicated and grew and caused epidemics. Baldwin would refuse to be responsible for what happened if a chemist was put in charge at Vigo.

"But you are responsible," Ditto said. "And you've got to continue being technically responsible for the whole operation."

"No, I can leave any time I want to," Baldwin said. "I'm still a civilian. I can pack my bags and go back to Wisconsin whenever I want. That was the deal from the very beginning."

Ditto backed down and Walter Nevius stayed on at Vigo, but the empty four-pound bomb casings from the Electromaster Corporation continued to pile up in the storehouse.

The day after Japan surrendered, the Vigo hierarchy formed a special steering committee to supervise the demobilization of the plant. The Vigo plant, at that point, had a total force of 1,500 Army, Navy, and enlisted personnel working on the base, a number that would soon be trimmed to 300. Meanwhile, excess stock left Vigo by the carload. Three boxcars of machine shop equipment, along with 20,000 empty Mk I bombs, went to Camp Detrick for storage. A stock of 765,000 unused detonators went to the Quartermaster Corps. One carload of sulfuric acid, 16,000 gallons of caustic, and 20,000 pounds of bleach went to the Army Service Forces Depot in Memphis, Tennessee. The supplies and equipment that nobody else wanted were given away as donations, to Emory University, Purdue University, the University of Michigan, and the U.S. Penitentiary at Terre Haute, among other places. Some 600 purchase orders for materials had to be canceled, including a $4 million contract with Electromaster for bombs.

The demobilization program fell behind schedule, too, and by the end of the year the Vigo plant had not yet been converted to inactive status. If and when it finally reached that point, the place was to be held in standby condition for a period of five years. What would happen after that, no one knew. In view of all the secret construction on the premises, the plant could not be sold or leased to private industry, the Army decided, and there was every indication that the Vigo plant, like Horn Island, would become another white elephant of the American biological warfare project.

Camp Detrick, too, came more or less to a standstill.

At the end of the war there had been some 200 separate projects under way at the post, everything from anthrax spore production to the mass cultivation of killer mosquitoes to research on the dissemination of plant diseases such as late blight of potatoes and brown spot of rice. At peak strength, 2,273 people were working at Camp Detrick, including 1,702 Army personnel, 562 Navy, and 9 civilians.

The camp had blossomed far beyond the Army's initial expectations. In the beginning, back in the early days of April 1943, the Chemical Warfare Service had set aside $1.25 million for the construction of the Detrick base. That would suffice, they had thought, for the required technical facilities, worker housing, and administration buildings. But just three months later, in July, the number had more than tripled, to $4.3 million. The increase covered additional research facilities and equipment, as well as the impact of all this new construction on the Frederick economy: whereas the basic labor rate had been forty-five cents an hour at the outset, the wage had risen to seventy-five cents an hour within three months.

The physical plant that finally arose at the foothills of the Catoctin Mountains bore little resemblance to the trim research facility that Ira Baldwin originally had in mind. By the war's end there was a small city inside the perimeter fence, a collection of more than 245 structures, including a hospital, fire house, laundry, chapel, theater, library, post exchange, and swimming pool, as well as several recreation halls. The swimming pool had gone up in record time, with volunteers shoveling out dirt by hand during their off hours. All this was in addition to the yellow-brick lab buildings, the pilot plants, the water, air, and steam-sterilization plants, plus the sewage sterilization system, animal incinerators, and housing for 5,000 workers. As of November 1, 1945, the final cost of Camp Detrick was $12,271,700.07, ten times more than the original estimate.

In its two and a half years of existence as a germ warfare research and development center, the place had run through amazing numbers of lab animals. Between August 1943 and December 1945, seventeen different species of animals had been utilized in experiments at Detrick. The final tally was: 598,604 white mice, 32,339 guinea pigs, 16,178 rats, 5,222 rabbits, 4,578 hamsters, 399 cotton rats, 225 frogs, 166 monkeys, 98 brown mice, 75 Wistar rats, 48 canaries, 34 dogs, 30 sheep, 25 ferrets, 11 cats, 5 pigs, and 2 roosters—a grand total of 658,039 individual specimens. Those not bred on the post's own animal farm had come from the Jackson Memorial Laboratory at Bar Harbor, Maine; the Bagg Research Laboratory in Westchester County, New York; and from the Army Medical College in Washington, D.C., all of which raised animals for laboratory use.

Most of those specimens were long since dead and gone. Between February 1944 and June 1945, the scientists autopsied some 4,000 lab animals at Detrick. They'd taken an average of three to five tissue samples from each specimen, made two or three slide sections from each sample, and generated a final total of 25,000 microscope slides.

They'd grown pathogens in the lab buildings by the test tube and by the flask, and in the pilot plants by the pound and by the ton. And they'd done a substantial amount of defensive work, although with not very much to show for it in the end. Researchers in several Camp Detrick divisions had worked on a variety of defenses against biological agents, for the purpose of protecting both the researchers themselves and the troops who might be exposed to those agents if and when they were used on the battlefield. The defensive measures included physical devices such as biological masks (chemical masks had been invented in World War I, and in 1942 Walt Disney had designed a Mickey Mouse gas mask for use by children), hoods, and protective clothing. This gear afforded variable amounts of protection, but the only surefire outfit was an impermeable rubber suit that was tolerable for short stints in the lab but was impractical for long-term use by foot soldiers.

The other type of defensive work was research on vaccines and toxoids, and after-exposure drug treatments. The Detrick scientists tried to develop immunizations for their star pathogen, anthrax, and for brucellosis, tularemia, and other diseases. Louis Pasteur had produced an effective anthrax vaccine for animals in 1881, but it was deemed far too dangerous for use by humans. No one else had created a safe vaccine in the interim, and the Detrick scientists were not successful at it either in the few years they had to work on the problem. There were experimental vaccines for brucellosis and tularemia, but they were unreliable and therefore could not be given to troops. Vaccine development was slow going, and it sometimes took ten weeks or more for results to show up in lab animals. Even then there was no guarantee that what worked for mice or guinea pigs would be transferable to humans.

Defensive research at Camp Detrick never met with as much success as offensive work, it being much easier to blow a pathogenic aerosol out of a bomb than it was to perfect ways of neutralizing the effects of a biological agent once it had entered the human body. Except for botulism, for which Pappenheimer had developed a toxoid in 1943, and drug therapies for certain diseases (anthrax responded well to streptomycin and penicillin), there were no acceptable defenses against most biological warfare agents by the end of the war.

Less than a week after V-J Day in September 1945, Detrick administrators cut work schedules by twenty-five percent and instructed the scientists to close out their projects within three months. Pending a plan for postwar operations, a handful of research projects would continue.

All that remained were the postmortems on the program itself.

"There has been a certain amount of duplication of effort in the several countries," David Henderson said. The Americans, he complained, "have attacked de novo problems which the British had brought to a satisfactory state of solution."

"The two big sour notes at Detrick," said Walter Nungester, a Detrick bacteriologist, "were the time and money wasted in the study of N and our slowness in accepting or even adequately trying unconventional modes of dissemination such as use of insects, refrigerated munitions, and other ideas."

Some of the scientists objected to the undue secrecy requirements placed on their work—especially the way it had throttled their publication opportunities and chances of advancement in the outside world. Still, all of it had been tolerable under the pressure of war.

With the war over and the threat gone, and with biological weapons never having been used on the battlefield by any of the combatants on either side, one might imagine that germ warfare research would quietly wither away and die. What actually happened was just the reverse.

part two

7

bad as U.S. Army Intelligence had been in assessing the true state of the German and Japanese germ warfare programs, they occasionally managed to come up with one or two items of useful information. In December 1944, one intelligence report told of a Japanese biological warfare research center in the hinterlands of Manchuria, and mentioned a certain "Maj-Gen Shiro Ishii."

Not until the spring of the following year did the Americans get further details. In a report dated April 6, 1945, and entitled "BW Information— Source: Captured Personnel and Material Branch," the War Department's Military Intelligence Service described what was apparently a major biological warfare research complex at Harbin, China:

> *Bacteriological Experimental Center (Saikin Kenkyu sho), Harbin:* Four Ps/W [prisoners of war] verified the existence of this agency, which is Army-controlled, but could not locate it on a map. Experiments are being carried on by Army biologists. There were no civilian officials connected with this center, which is commanded by Maj-Gen Ishii, Shiro. Nature of the types of experiments being carried on here is extremely secret and their findings were never published for general assimilation.

The account named several other such germ warfare centers in China and Japan, including one in Tokyo that was also commanded by Shiro Ishii.

Gijutsu Sho, Tokyo. Believed that this center is, also, engaged in biological experiments under Army supervision. Exact location unknown. In 1941 or 1942, Maj-Gen Ishii, Shiro, received a medal for the development of a water purification machine. This was announced in the newspapers.

To the officials of the U.S. Army Chemical Warfare Service, these bits of news were highly intriguing. The Japanese, they knew, had been conducting plague experiments in China by airplane in 1940, at a time when the orthodox CWS viewpoint had been that germ warfare was an impractical dream. If, as it now appeared, the Japanese had continued to investigate biological weapons throughout the course of the war, they might have made substantial progress. In that case, the Army decided that it ought to find out what the Japanese had discovered, and the sooner the better, before any competitor—the Russians, for example—managed to do so. It would be especially interesting, above all, to find and interrogate this "Maj-Gen Ishii, Shiro." And so in September 1945, within a month of the Japanese surrender, a Camp Detrick scientist by the name of Murray Sanders arrived in Tokyo to question the Japanese biological warfare scientists and collect data.

Murray Sanders was a Detrick microbiologist who'd taught at Columbia University's College of Physicians and Surgeons before the war. He was young and dashing and wore a mustache, and he was always eager to please.

Sanders had entered the Army in 1943 and was assigned to Camp Detrick, where he took up the post of section chief and worked with botulinum and staphylococcal toxins, among other things. In December 1944 he and Arvo Thompson, who was then Ira Baldwin's executive assistant, had examined the Japanese paper balloon that had been found floating off the California coast. Neither they nor later investigators ever found any suspicious bacteria on the objects which, in the end, turned out to be balloon bombs containing conventional explosives. On May 5, 1945, one of the balloons had landed on Gearhart Mountain near Bly, Oregon, killing a family of six who had come across the object while hiking.

When Murray Sanders arrived in Japan, he was met at the dock by Ryoichi Naito, a Japanese physician who spoke decent English and announced that he would act as Sanders's interpreter. Although the name meant nothing to Sanders, this was the same Ryoichi Naito who, more than five years earlier, in February 1939, had been in New York City trying to get yellow fever virus samples out of the Rockefeller Institute for Medical Research. Since that time Naito had worked in Ishii's biological warfare unit at the Japanese Army Medical College in Tokyo, trying to make a weapon out of fugu toxin, the poison found in blowfish livers. His efforts had been interrupted by American B-29 bombing raids over Tokyo in the latter part of 1944, but he continued to work in various Japanese germ warfare installations.

And here he was now escorting Camp Detrick's advance man, Murray Sanders, all over the city.

Sanders set himself up in the Dai-Ichi building, home of the Supreme Allied Headquarters, where he received a procession of Japanese biologists and military men who were brought in for questioning. He interrogated not only Naito himself but also Ishii's longtime friend Tomosada Masuda, Army Surgeon General Hiroshi Kambayashi, biological bomb expert Jun'ichi Kaneko, and Army technical expert Seichi Niizuma, among others, over the course of a ten-week stay in Japan. Most of his informants spoke easily enough about defensive germ warfare research but tended to clam up the moment offensive work was mentioned. But Sanders kept on with his questioning, and through a combination of charm, subterfuge, and grim persistence, managed to learn a considerable amount.

Naito told him how the Japanese biological warfare effort had gotten started in 1932, when Shiro Ishii, then a major in the Kwantung Army, returned to Japan after a tour of Europe. He'd been convinced by the 1925 Geneva Convention banning germ warfare that bacteria must have great potential as weapons; otherwise they would not have been prohibited. Ishii tried for many years to get research money but did not succeed until 1937, when the War Ministry provided funds for the Ping Fan institute.

Sanders, in his written report on the interview, at this point added a notation that "the BW installation is actually located one hour by motor, due south of Harbin, in the vicinity of the small village of Ping Fan, a few miles east of the South-Manchuria Railroad between Harbin and Hsinkiang." Neither Sanders nor any other U.S. Army investigator had ever been to Ping Fan, which was by that time under Russian control.

Naito said that the Ping Fan institute had produced enormous quantities of vaccine for defensive purposes—some twenty-one million doses in a single year. Still, the site's main mission had been to develop bacterial agents into practical weapons, he said. Researchers there had invented and tested bacterial bombs by the hundreds if not thousands, dropping them from planes onto the experimental field adjacent to the institute and elsewhere.

There was the "Ha" bomb, for example, an anthrax weapon designed to kill by shrapnel penetration. The bomb was two feet long by six inches in diameter and contained an inner core of explosive powder surrounded by a middle layer of anthrax suspension and then an outer layer of steel shrapnel pellets. When the Ha bomb exploded, the anthrax coated the shrapnel, then the pellets shot out in all directions and sank into the skin of the target.

The Japanese had tested the Ha bomb on 100 horses and 500 sheep, and it had worked well enough, many of the animals being killed by a single contaminated shrapnel pellet. Still, the bomb had several limitations. When it fell into holes or depressions, the pellets flew out harmlessly into the ground. Also, the bomb's payload was small, for it held only about a pint or so of the anthrax suspension. Much of the anthrax, moreover, was destroyed by the large explosive charge necessary to fragment the bomb and scatter the shrapnel. The scientists were convinced it was possible to do better than this, and so they came up with a second device, the "Uji" bomb.

The Uji bomb had been designed by a civil engineer working at Ping Fan, a man by the name of Yamaguchi. Yamaguchi's major innovation was to make the bomb walls out of porcelain rather than steel. The porcelain shattered far more easily and so a much smaller explosive charge was sufficient to burst the casing, which meant that fewer bacteria were killed by the heat and shock of the blast.

That, thought Murray Sanders, was a highly ingenious concept.

The Uji bombs, Ryoichi Naito went on, were about the same size as the Ha bombs, but their porcelain walls were less than half an inch thick, and the explosive was on the *outside* of the bomb, a four-yard length of Prima-cord cemented into grooves that crisscrossed the bomb's outer surface. Since there was neither shrapnel shot nor explosive on the inside, the whole inner cavity could be filled to the brim with bacterial fluid, to a capacity of ten quarts. The device weighed eighty pounds when fully loaded, and it was stabilized in the air by celluloid tail fins that burned up in the blast so as to leave no traces that a bomb had fallen.

Some 500 Uji bombs were tested at Ping Fan, with both simulants and live pathogens, and although the devices worked quite well, they, too, had their deficiencies. For one thing, the porcelain casings were temperamental and difficult to manufacture. They were fragile and sometimes cracked prior to detonation. And, even though they were designed to explode in the air so as to produce the widest possible pathogen cloud, the height of the explosion couldn't always be controlled precisely over irregular terrain.

That problem was overcome by a third design, one which Murray Sanders regarded as by far the most ingenious of all. This was the so-called "mother and daughter" bomb, designed at Ping Fan by a Lieutenant Gondo, an aeronautical engineer with a special interest in radios and remote control.

His invention consisted of a large "mother" bomb containing a sending radio, but no bacteria, and a cluster of "daughter" bombs, each containing a receiving radio, an explosive charge, and hot agent. The plan was to release the mother bomb just before reaching the target, then drop the daughter units out a short time later. The radio system was designed so that when the mother unit contacted the ground, it transmitted a signal to the daughter bombs that caused them to explode in the air at a designated height over the terrain, thereby setting up a pathogen cloud of perfect shape, size, and density.

That, thought Sanders, was a damned clever device.

On Tuesday, October 9, and again on Thursday the 11th, Murray Sanders interrogated Tomosada Masuda, possibly the number-two man in the whole Japanese biological warfare operation. Masuda had been boyhood friends with Shiro Ishii, had worked in Ishii's Water Purification Unit beginning in 1937, and by 1939 had become the director of a new germ warfare facility in Nanking. He was an accomplished scientist, having gotten a medical degree in Kyoto in 1926 and then done postgraduate work in both France and Germany.

Masuda, who appeared to be sick and was supposedly suffering from malaria, told Sanders that the Japanese germ warfare program had begun in 1935 when a group of Russian spies had crossed over into Japanese-controlled territory in China. Five of the spies had been apprehended carrying glass bottles and ampules which, when examined, revealed the presence of cholera bacteria, anthrax spores, and dysentery organisms. "I personally saw the anthrax organisms," he claimed. It was this incident, said Masuda, that had motivated all of Ishii's bacterial warfare work.

Shortly thereafter, Masuda was involved in the germ warfare business himself, having been appointed by Ishii to two posts in the *Boeki Kyusui Bu*, the Anti-Epidemic Water Supply Department, first in Dairen, China, then in Nanking. In 1942, while he was a professor of bacteriology at the Army Medical College in Tokyo, Masuda had written a paper, "Bacteriological Warfare," in which he'd laid out the state of the art as it was known at the time. He summarized previous research, including the German simulant attacks on the Paris and London subway systems. He claimed that without any doubt "vectors and rodents can be employed in spreading diseases." And, in a section that surveyed foreign publications for lists of offensive bacteria, he duly recorded the suggestions that had been made by Leon Fox:

> Fox   (Maj, US Med Corps—1933)
> 1st Group—typhoid, para-typhoid, dysentery,
> cholera, typhus, plague; 2nd Group—gas
> gangrene, tetanus, anthrax; 3rd Group—
> botulinum toxin.

Over the course of two days, Masuda, despite his illness, plied Sanders with vast quantities of technical data as to the types of organisms cultivated at Ping Fan, test results, and conclusions. He admitted to having members of his staff poison some 1,000 water wells in China and then testing them for comparative bacterial survival rates. His men filled bottles with typhoid or paratyphoid germs and gave them to soldiers who dropped them into wells. Such a strategy was ineffective, Masuda decided, for all the organisms had died within days.

Masuda talked about the open-air test grids at Ping Fan and described the brightly colored chemical dyes that they'd added to the bomb mixtures to make the bacterial clouds more clearly visible. He also discussed the matter of accidental deaths among the test personnel. One man died from anthrax inhalation after he'd cut the grass at the experimental site on the day following a field trial. There were five other anthrax deaths during field tests, but Masuda did not say how the deaths occurred and Sanders did not press him on the point. In 1937, said Masuda, there had been two deaths from glanders (a respiratory disease transmitted to humans from horses), "due to carelessness on the part of the laboratory workers," and "in 1944, there were two plague deaths as a result of field trials."

Later, Sanders interviewed several other officials who had been involved with the Japanese germ warfare program, and they'd added corroborating data and new information. But Sanders couldn't shake the feeling that there was always something missing from their accounts, that some minor or major detail was consistently being withheld.

At the end of October, Sanders wrote a fourteen-page report laying out his overall findings and conclusions, an account that gave the first clear picture of the Japanese biological warfare program. This was his summary:

SECRET

Summary
BIOLOGICAL WARFARE (BW)

1. Responsible officers of both the Army and Navy have freely admitted to an interest in defensive BW.
2. Naval officers maintained that offensive BW had not been investigated.
3. Information has been obtained that from 1936 to 1945 the Japanese Army fostered offensive BW, probably on a large scale. This was apparently done without the knowledge (and possibly contrary to the wishes) of the Emperor. If this was the case, reluctance to give information relative to offensive BW is partially explained.
4. BW seems to have been largely a military activity, with civilian talent excluded in all but minor roles.
5. The initial stimulus for Japanese participation in BW seems to have been twofold:
   a. The influence of Lt Gen Shiro Ishii.
   b. The conviction that the Russians had practiced BW in Manchuria in 1935, and that they might use it again. (The Chinese were similarly accused.)
6. The principal BW center was situated in Ping Fan, near Harbin, Manchuria. This was a large self-sufficient installation with a garrison of 3,000 in 1939–1940. (Reduced to 1,500 in 1945.)

7. Intensive efforts were expended to develop BW into a practical weapon, at least eight types of special bombs being tested for large-scale dissemination of bacteria.

8. The most thoroughly investigated munition was the Uji type 50 bomb. More than 2,000 of these bombs were used in field trials.

9. Employing static explosion techniques and drop tests from planes, approximately 4,000 bombs were used in field trials at Ping Fan.

10. By 1939, definite progress had been made, but the Japanese at no time were in a position to use BW as a weapon. However, their advances in certain bomb types was such as to warrant the closest scrutiny of the Japanese work.

11. Japanese offensive BW was characterized by a curious mixture of foresight, energy, ingenuity, and, at the same time, lack of imagination with surprisingly amateurish approaches to some aspects of the work.

The Japanese program, it turned out, mirrored the American germ warfare project effort to a surprising degree. Both programs had studied the same gamut of organisms, chief among which were plague, anthrax, glanders, typhus, dysentery, and cholera. Researchers in both countries sought the same types of data, concerning the best routes of infection, the optimum particle size, the minimum infectious and lethal doses, and the ideal size, shape, density, and persistence rates of pathogen clouds. To make the clouds more visible, both the Japanese and the Americans had used the identical chemical dye, methylene blue. Both had shellacked the interior surfaces of their bombs to prevent corrosion. Both had used the same two simulant organisms, *Serratia marcescens* and *Bacillus globigii*.

There were also some differences. The Japanese had not shown much interest in either viruses or rickettsiae, a lapse that prompted Murray Sanders to speak dismissively of their "limited imagination insofar as the virus-rickettsial agents were concerned. Why this group was not even considered in the selection of agents is not clear." Also, the Japanese had never used cloud chambers at any point. Chambers were unnecessary, Masuda had explained, since all the test work could be carried on directly in the field.

And they were extremely slow to put together anything resembling a weapons mass-production line. Whereas the Americans had gone from program inception to the Vigo plant, such as it was, in about eighteen months, the Japanese had been in the bacteriological warfare business for about ten years, from 1936 to 1945, but so far as Sanders could discover they'd never gotten to the point of producing an actual inventory of working bombs.

Sanders put all this in his technical report and capped it with a supplement containing organizational charts of the Japanese germ warfare units, maps of the Harbin area, plans of the grounds and buildings at Ping Fan, plus photographs and line drawings of the various biological bomb types tested there. He added an appendix giving summaries of all the interrogations that he'd conducted, plus stenographic transcriptions of a few selected interviews. The main item missing from all this was an interview with the fabled Shiro Ishii himself.

"All informants are agreed," wrote Murray Sanders, "that this individual was the compelling force behind the scenes throughout the period of Japanese investigations in the field of biological warfare."

Ishii, however, seemed to have vanished.

In October 1945, while Murray Sanders was still in Japan, George W. Merck, the civilian boss of the United States biological warfare effort, submitted a report to the secretary of war outlining the nation's activities in the field. The document, which came to be known as "the Merck Report," was originally classified as Secret, but on January 3, 1946, the War Department's Bureau of Public Relations released a sanitized version to the press, and the country learned for the first time of the germ warfare program that had been conducted under its nose for the previous three years.

The expurgated report disclosed and withheld information in roughly equal amounts. It explained how the War Department had launched germ warfare research for defensive purposes, and for the further aim of retaliating in kind against a biological attack. It revealed that American scientists had studied "all known pathogenic agents" for their possible use as offensive weapons and that the most promising agents "were assigned to various university and private research laboratories for intensive experimentation."

The report did not identify any university or private lab by name, nor did it say how many had been involved. The fact was, however, that the War Research Service had paid biologists at more than two dozen colleges, universities, and both private and public laboratories across the country to investigate defenses against or the offensive possibilities of assorted microbial agents. Scientists at the National Institute of Health had studied both cholera ("HO") and typhus ("YE"). René Dubos at the Harvard Medical School had investigated dysentery ("Y"), while Cornell had worked on anthrax ("N"), the University of Cincinnati on tularemia ("UL"), Michigan State College on brucellosis ("US"), Northwestern on mussel toxins ("SS"), Notre Dame on rickettsiae ("RI"), and so on down the long list.

One experiment that the Merck report made no mention of at all was the U.S. Navy's use of fifty San Quentin convicts as human guinea pigs in bubonic plague ("LE") tests, work that had been conducted by the Naval Research Unit at the University of California, Berkeley. When the Navy's scientists had sent out a request for volunteers, prison warden Clinton Duffy made an announcement over the "Gray Network," San Quentin's public address system, and was immediately besieged by 200 men, prisoners serving time for everything from forgery and car theft to murder. After personal interviews and physical fitness examinations at the prison hospital, the fifty winners were selected and were later injected with the plague organism. None became seriously ill, although several of the men developed sore arms and headaches two days after the injections.

All this had remained a closely guarded secret, but the day after the War Department released the Merck report the Navy itself held a press conference and told much of the story. Next day there was a front-page story in the *Oakland Tribune*, headlined "Convicts Aid Germ Battle." There was no public uproar then as there might have been in later times: the men, after all, had volunteered for the project, and they had arguably helped the war effort.

The Merck report spoke elliptically of a pilot plant center "in Maryland," a field test site "in Mississippi," and a large-scale production facility that the Army had set up "in Indiana." The report was not more forthcoming with names, places, and concrete details because, at least in George Merck's view, the threat of biological warfare still hung over the United States. "It is important to note that, unlike the development of the atomic bomb and other secret weapons during the war, the development of agents for biological warfare is possible in many countries, large and small, without vast expenditures of money or the construction of huge production facilities," he said. "It is clear that the development of biological warfare could very well proceed in many countries, perhaps under the guise of legitimate medical or bacteriological research."

Nobody knew what new and horrific evils might even then be under development in unknown countries, weird forms of biological aggression that the United States might have to retaliate against in the future. For that reason, Merck advised that the country's germ warfare research program continue on pretty much as before.

As indeed it did. Although Horn Island was defunct and the Vigo plant had been mothballed, basic research into offensive pathogens kept going at Camp Detrick.

The most pressing need among the scientists who remained there was for a way of testing agent-filled bombs without traveling clear across the country to do it. The Detrick researchers wanted to be able to explode actual munitions loaded with hot agents over ranks of live test subjects, but at the start of 1946 the only place in the country where this was possible was at the Granite Peak Test Station, part of Dugway Proving Ground, in Utah.

The Army had set up Dugway in January 1942 as a chemical weapons test center. It was located on a vast flat plain about the size of Rhode Island in the Utah desert some eighty miles west of Salt Lake City. With few humans in the immediate vicinity, and not that many animals, it was ideal for experimenting with the latest advances in flamethrowers, incendiary bombs, smoke munitions, and whatever else. In 1944, soon after it became clear that Horn Island was going to have severely restricted usefulness as far as live pathogen testing was concerned, and that in fact it would be an abject failure, the Army established a separate 250-square-mile tract near Granite Peak, a large shelf of vertical rock in the area, and reserved it for biological operations.

As they had at Detrick, Horn Island, and Vigo, construction engineers now came to Granite Peak and fitted the place out with the requisite facilities, this time including twenty-two miles of hard-surfaced roads, living quarters for the test personnel, a water supply system, sewers and septic tanks, power generators, and an airplane landing strip. The biological installation itself consisted of test grid, lab, and the inevitable incinerator for used-up animal carcasses. The site that became known as the Granite Peak Installation was finally completed at the end of January 1945, seven months before the end of the war, at a total cost of $1,343,334.

Camp Detrick scientists came to Granite Peak and tried out a few new agent-munition combinations, such as a ninety-one-pound air-burst bomb containing "VKA," vegetable killer acid, for possible use against Japanese rice crops. But this was too late in the war to have had any appreciable impact on the final outcome, and the Army shut down the post at the war's end. Although the site was supposed to remain available for test purposes during a few months of every year, Utah was still a long way for the Camp Detrick researchers to go every time they wanted to test the effects of a new live agent or the latest modification to a biological bomb.

There was a similar site for field trials in Canada, the Suffield Experimental Station, near Medicine Hat in Alberta. This was a huge expanse of denuded prairie that the Canadians had set aside in 1941 for chemical weapons tests. Later, in 1944, when biological materials became available in quantity, the Canadian scientists had marked out a separate section, Area E, thirty miles north of the main base, for biological weapons tests. Crews from Detrick had gone up there, too, mainly with four-pound Mk I bombs loaded with suspensions of *Brucella suis* ("US"), and had fired them over stands of boxed guinea pigs and other animals, with uneven results. There had been one especially puzzling trial in which the normally highly virulent "S1" strain of the brucellosis organism was disseminated over an array of thirty-five six-week-old hogs, and not a single animal became infected after exposure, even though the sample itself, which was six months old, was found to be viable when it was sent back to Detrick and analyzed in the lab.

In a later technical meeting at Suffield, Dr. Leroy Fothergill, the new scientific director of Camp Detrick, reported that by the time of the lab tests, some of the *Brucella* suspension had clotted together into large sticky lumps, and he guessed that this may have adversely affected the particle size of the agent during the field trials. But he acknowledged that this was mere conjecture on his part and that he really could not explain why the agent had not performed as it should have.

Such puzzles were only part of the normal progress of science, the standard birth pangs of a new technology. Still, the Suffield *Brucella* failure emphasized the fact that if biological weapons were ever to play a role in the U.S. war machine, then germ warfare would have to be raised from the level of mere practical lore into something approaching a true science. You had to be able to predict the mortality rates of different types of pathogen clouds on a range of exposed animals and then extrapolate the likely mortality rates of those same agent clouds on human beings.

That would require huge volumes of empirical data, data that could come only from a long series of hot agent trials over live animals. It would be inconvenient in the extreme if the Detrick scientists had to make a separate jaunt into the Utah desert or the Canadian outback whenever they wanted to try out something new. What they needed was something closer to home, such as an enclosure at Detrick in which they could fire off bombs loaded with live pathogen and then let the resulting aerosol pass over the test specimens.

What they needed, in other words, was a bombproof exposure chamber. Soon they would get one.

On November 10, 1945, while Detrick investigator Murray Sanders was en route back to Maryland, the mayor of Chiyoda-Mura, Shiro Ishii's hometown, announced that Lieutenant-General Ishii had been shot to death.

No further details were made available other than that Ishii was survived by his elder brothers Takeo and Mitsuo (both of whom had worked alongside him at Ping Fan, Takeo as the prison commander, Mitsuo as supervisor of the animal house), his wife, Kiyoko, and his daughter Harumi, as well as five other children. He was fifty-three years old at the time of his unfortunate assassination at the hands of person or persons unknown.

Ishii had been born on June 25, 1892, at 1382 Osato Street, Chiyoda-mura, in Chiba Prefecture, a few miles southeast of Tokyo. The residence was in fact a country villa, a sprawling collection of low buildings, tall trees, fruit orchards, and groves of green bamboo. The Ishii family, a clan of wealthy landowners, had lived there from the middle of the nineteenth century onward.

A short distance from the main house, in a shady spot located within a copse of trees at the top of a small hill, was the family cemetery. Ishii's eldest brother, Torao, who had died in the Russo-Japanese War, was already buried there.

In mid-November, an elaborate funeral, complete with priests, mourners, incense, prayers, and offerings, took place in the village of Chiyoda-Mura. There was a small memorial service after which the procession wended its way through the village, then turned down Osato Street in the direction of the Ishii compound. Once inside the gates, the mourners climbed the hill to the graveyard and disappeared from view amid much lamentation and sadness over the untimely death of "the Honorable Ishii."

8

hree weeks after Shiro Ishii's funeral, the Tokyo headquarters of the U.S. Army Counter Intelligence Corps sent out a memorandum to selected intelligence units in the area, as well as to the office of the war crimes trials in the Pacific:

CONFIDENTIAL

HEADQUARTERS

COUNTER INTELLIGENCE CORPS
METROPOLITAN UNIT NUMBER 80
GHQ AFPAC

APO 500

3 December 1945

SUBJECT: ISHII, Dr. Shiro
Chioda-mura, Yamatake-gun
Chiba Prefecture

Summary of Information:

The following information was obtained from Confidential Informant 80-11 of this office, during the course of an investigation:

On 10 November 1945, SUBJECT, allegedly a large landowner in Chiba Prefecture and a former Lt. General of the Army Surgeon's Corps of the Japanese Army, was proclaimed dead and his funeral observed in Chioda-mura, Yamatake-gun, Chiba. This death is alleged false by Informant 80-11, and it is claimed that ISHII has gone underground with the aid of the Village Headman of Chioda-mura, and intends to carry on anti-democratic activities.

It is claimed that SUBJECT was the Commanding Officer of the Ishii Detachment of the Surgeon's Office during the war; that he had his assistants inject bubonic plague bacilli into the bodies of some Chinese in Harbin, China and some Americans in Mukden, China as an experiment; that the subject carried out a similar experiment in Canton, China, and that as a result of his carelessness there the bubonic plague ravaged the city.

Evaluation

—of source	—of information
B [reliable]	3 [probably true]

A second report soon arrived claiming that Ishii was hiding out in some unknown mountain retreat. Then a third announced that

Ishii returned and is said to have reported to Prof. Tatsuo Ishikawa, Professor of Pathology at Kanazawa Medical College, near Kanazawa City.

Ishii is said to be hiding somewhere near or in Kanazawa.

On January 9, 1946, the American occupation forces in Japan ordered the Imperial Japanese Government to apprehend Shiro Ishii in Kanazawa, bring him to Tokyo, and hand him over to the Supreme Commander of the Allied Powers.

On January 11, two days later, Lieutenant Colonel Arvo T. Thompson, of the U.S. Army Veterinary Corps, arrived in Japan from Camp Detrick. Thompson, at this point, was an old hand at the American biological warfare game, having been assigned to the Chemical Warfare Service in Washington, D.C., on March 30, 1942, almost a full year before Ira Baldwin drove out to Frederick and found it to be a suitable location for a germ warfare research center. Later, Thompson became the technical aide to George W. Merck at the National Academy of Sciences. In February 1944, he was transferred to Camp Detrick and he and his wife moved out to Frederick, where he became executive officer to Ira Baldwin.

Arvo Thompson was tall and dapper and wore a pencil mustache. By the time he left for Tokyo shortly after Christmas 1945, he was thoroughly familiar with Murray Sanders's report of his interrogations of the Japanese germ warfare scientists. What Sanders had accomplished, however, was only the merest beginning. Sanders had never even met Ishii, much less questioned him about his role in the Japanese program. Ishii, however, was very much the star of the show, as all later intelligence reports had confirmed. One of the Army's "very reliable" Tokyo informants had referred to Ishii as "the germ man," and had said that "the whole career of this adventurous surgeon starts with bacteria and ends with bacteria." If, as a number of reports claimed, Ishii had performed germ warfare experiments on living human subjects, then as one U.S. Army Intelligence officer had put it, "the results of these experiments are of the highest intelligence value." Conceivably, they were also of the highest value for the future of the American germ warfare project, since they would embody data as to the effect of pathogenic agents on their intended final destination, human beings.

Clearly, Ishii was the major attraction. The jackpot.

When Thompson left for Japan, however, neither Ishii's precise whereabouts nor even whether he was dead or alive was known for sure. But a week after Arvo Thompson arrived in Tokyo, special agents of the U.S. Army Counter Intelligence Corps had finally put their hands on Ishii. He had been hiding out not in the mountains or in Kanazawa, but at the family estate in Chiyoda-mura, where he was living quietly with his wife and daughter.

He was unwell, suffering from chronic gall bladder inflammation and dysentery—or so he claimed. But on January 17, the Japanese government, in response to a formal American request, forcibly returned Ishii to Tokyo and delivered him to SCAP, the Supreme Commander of the Allied Powers.

On January 22, 1946, a Tuesday, Arvo T. Thompson, D.V.M., paid his first formal call on Shiro Ishii, M.D. Thompson was accompanied by his interpreter, Second Lieutenant Francis M. Ellis, and by Lieutenant Colonel D. S. Tait, of U.S. Army Intelligence.

Ishii's home was on a secluded and narrow back street in the Wakamatsu-cho section of Tokyo. Ishii, attended by his daughter Harumi, seemed visibly unwell. The mustache he used to wear was gone. He looked thin and frail and did not fully fill out the clothes he had on. But both he and his daughter received Thompson and the other Americans politely and treated them as honored guests.

Thompson interrogated Shiro Ishii, in person and by means of question-naires, over the course of the next four weeks. He was not an aggressive interviewer, however, and seldom challenged or disputed Ishii's replies, no matter how dubious they might have seemed at the time. Ishii denied, for example, that he or anyone else had ever dropped plague bacteria into China.

Q. We have heard from Chinese sources that plague was started in Changteh, China, in 1941, by airplanes flying over and drop-ping plague material and a plague resulted. Do you know any-thing about that?

A. No. It is impossible from a scientific point of view to drop plague bacteria from airplanes.

Q. But rats, rags, and bits of cotton infected with plague were dropped and later picked up by the Chinese and that was how it was to have started.

A. If you drop them from airplanes they will die. There is no chance of a human being catching plague as a result of drop-ping organisms from an airplane.

So worried had he been about inciting a bubonic plague epidemic, said Ishii, that he hadn't even allowed any field tests with the agent.

Q. What field tests were made with the plague organism?

A. Due to the danger of it, there were no field tests with that organism. There were a great many field mice in Manchuria and it would have been dangerous to conduct field experi-ments with plague because the field mice would very easily carry the organisms and start an epidemic. We conducted experiments with plague in the laboratory.

Q. What kind of experiments?

A. We put rats in cages inside the room and sprayed the whole room with plague bacteria. This was to determine how the rats became infected, whether through the eyes, nose, mouth, or through the skin.

Q. What did you find out?

A. The results were not too favorable. We usually got ten percent infection.

Q. By which way?

A. Through the nose. Also, through an open wound. Animals were shaved and it was found that they would become infected through the microscopic abrasions caused by the shaving. We found that the lymph nodes became inflamed. That is how we knew if the animal had been infected.

Q. Was this spray test conducted in a special chamber or in an ordinary room?

A. It was not a special chamber. The windows were double-plated and paper was put all over the walls. The room was made as airtight as possible. Human beings did not enter the room. They conducted the test from an outside corridor.

Q. Was there not any danger in handling the animals after the experiment? And also was it not dangerous because of the bacteria still being in the air?

A. After the experiment, we sprayed formalin in the room and did not enter it for one day.

Q. How were you protected while handling the animals?

A. We wore protective clothing, masks, and rubber shoes. Before we touched the animals, we put the cages, animals and all, into a solution of creosol.

Q. Did you have any accidents in the laboratory or nearby as a result of the experiments?

A. Yes. A person who handled the animals after the experiment got infected and died.

Thompson asked Ishii about the Uji bomb.

Q. The Uji-type bomb was specially made for BW. Who made them?

A. The old laboratory in Harbin, which manufactured the water filters, had the facilities for baking the porcelain, and the same equipment was used for making the bombs as was used for the manufacture of the water filters. Most trouble I had was trying to get technicians for baking porcelain. The technicians occupied a separate room in the factory and they made a small inner factory for the manufacture of the porcelain bombs.

Q. Why was not the assistance of regular ordnance officers used in making the bombs?

A. In ordnance they deal only with iron bombs. They would not take the responsibility for making porcelain bombs. We could manufacture the bombs as we pleased and make any changes or improvements which we felt were necessary.

He asked about the balloon bombs.

Q. What work was done at Ping Fan Institute with the trans-oceanic balloon?

A. None. I think it was carried out in a scientific laboratory in Japan.

Q. Did you have any connection with the work?

A. None.

Over the course of time the exchanges came to be pleasant and friendly. Ishii's daughter Harumi often served tea, and she and her mother, Kiyoko, sometimes offered complete meals to Thompson, Ellis, and the other American servicemen who occasionally attended the questioning. The Ishiis set a fine table, with long-stemmed wine glasses, water goblets, and elegant displays of Japanese flowers.

For the benefit of Thompson and the others, Ishii at one point demonstrated the use of his celebrated water purification device. Another time, with the help of technicians from a nearby military hospital, he showed off the bacterial culture cabinet that he'd invented for the purpose of making vaccines and later had converted to the growth of bacteria for field trials.

The culture cabinet was an aluminum box about the size and shape of a microwave oven. It held a series of trays that could be filled with bacterial growth medium, and Ishii now poured out some melted agar into the trays and inoculated each one by swabbing the surface with a sample of the bacteria to be grown. Then he trotted out a second cabinet that he'd previously inoculated with *E. coli* bacteria and showed how to harvest the new growth by scraping the microbes off the top with a small metal rake.

Although there had been no communication between the Japanese and Canadian germ warfare scientists, this system was virtually identical to the one that the Canadians had used to culture anthrax bacteria on Grosse Ile.

Ishii seemed proud of his invention and the Americans were politely impressed by the appliance—insignificant and backward though it was in comparison to the 10,000-gallon fermenter in Pilot Plant 2 at Camp Detrick, to say nothing of the twelve separate 20,000-gallon fermenter tanks at the Vigo plant.

There is no record that Arvo Thompson ever asked Ishii about his alleged use of human guinea pigs. Ishii, for his part, took pains to defuse rumors that anything untoward or indecorous had ever happened at the biological warfare institute that he had commanded at Ping Fan.

A. A lot of men in my unit, and others who do not know anything about it, have been spreading rumors to the effect that some secret work has been carried on in BW, and they have gone so far as saying an attack was planned by my unit and that a lot of bacteria were being produced, large quantities of bombs manufactured, airplanes gathered for that purpose. I want you to have a clear understanding that this is false.

There had been, Ishii conceded, a lot of bomb research going on at Ping Fan. There was the "I" bomb, for example, which Thompson later described as "perhaps the first munition developed for the dissemination of a bacterial liquid payload." When it hit the ground nose first, the bacteria sprayed out from the tail. The Japanese had first tested it in 1937 and had learned that the bomb often went so deep into the ground on impact that all the bacteria was effectively buried. To overcome this they developed a second bomb, the "Ro" bomb.

The Ro bomb was similar to the I bomb except that on contact with the ground, the front part of the bomb exploded and threw the bomb's payload section ten to fifteen meters into the air. Then the payload section itself exploded, blowing the contents out in a fine spray. There was also a "Ga" bomb, similar to the porcelain Uji bomb except that the Ga bomb was made of glass. All these been had tested at Ping Fan, on various types of large animals including oxen, horses, goats, and sheep.

Before he left Tokyo, Thompson accepted and took away with him the organizational charts of the Japanese biological warfare command, maps of the Ping Fan institute, sketches of bombs, and other documents that Ishii had provided him in grand profusion.

Later, Arvo Thompson interrogated some of the pilots who had dropped these bombs on the Ping Fan test range, and they substantially verified Ishii's account. Once, on February 20, in an interview with Lieutenant Colonel Yoshitaka Sasaki, a surgeon, Thompson asked about rumors of experimentation on humans.

> **Q.** Have you heard whether the Chinese or POW's were used in BW experiments?
> **A.** I do not know.

And that was that. Apparently, neither Thompson himself nor any of the other Army Intelligence officers involved in the case could bring themselves to believe that Ishii had permitted any sort of human experimentation—claims that anyway had stemmed from members of the Japanese Communist Party.

At the end of his two-month stay in Japan, Arvo Thompson came back to Camp Detrick and wrote up his formal "Report on Japanese Biological Warfare (BW) Activities," a document that ran to some fifty pages in length. In it were a couple of disclosures that ran counter to claims made by Murray Sanders: viruses and rickettsiae had indeed been experimented with at Ping Fan. On the other hand, "Ishii denied the existence of a 'mother and daughter' radio bomb."

For all his diffidence as an interrogator, Thompson did not take the whole of Ishii's testimony at face value. "On the subject of BW research," Thompson wrote, "Ishii's replies to questions were guarded, concise, and often evasive." Nor could he accept Ishii's claims that all documentary evidence, including the bomb prototypes, had been lost in the final days of the war. The Ping Fan institute and everything in it, Ishii had told him, had been intentionally destroyed prior to the Russian advance into the Harbin area. However, "the detailed bomb sketches and other technical information obtained from Ishii indicate an amazing familiarity with detailed technical data. It leads one to question the contention that all records pertaining to BW research and development were destroyed."

Those records, Arvo Thompson thought, could be of crucial importance to the future progress of biological warfare research in the United States. Someone should be sent to collect them.

■   ■   ■

At his home in Tokyo, Shiro Ishii began to receive anonymous letters, phone calls, and, on occasion, hand-delivered messages from former associates. All of them contained demands for money, and many of them were backed by veiled threats.

On June 3, 1946, a woman dressed in a streetcar worker's uniform showed up at Ishii's house at 77 Wakamatsu-cho with an envelope for him. Two men, one about thirty, the other in his early twenties, had asked her to hand deliver it. The envelope, which bore Ishii's name and address, was marked "Absolutely Confidential." The letter inside it said:

> 3 June
> Matsunosuke Masegawa
> Masao Morii
> Tamio Yoshida

To   Your Excellency Shiro Ishii
     Former Lt Gen, Army Med Corps

Dear Sir:

You must be surprised to receive this badly scribbled rude letter so unexpectedly. We were one of your subordinates (you were the CO) at Nanking. After the termination of the war, we came back to Japan, but the defeated Japan was not very cordial in seeing us back. Our homes were burnt and our wives and children were dead, and though we didn't have a cent to our names, we did manage to rehabilitate ourselves but the waves of inflation have finally subdued us, we are experiencing difficulties in obtaining our tomorrow's food. Because of our hardships, we are about to fall into committing wicked things, but by all means we want our thoughtful commanding officer to rescue us.

While we were at Nanking, we were ordered to carry out some gruesome work, and we did perform our duties faithfully. It must have been a difficult task to bury all those materials after the war was over.

Because of our present hardships, we thought of dying more than just once, but when we thought about it, we realized that if we have enough courage to die, then we certainly should be able to overcome them and accomplish anything.

Please, we beg you, our commanding officer, that you loan us, the unfortunate ones, as our rehabilitation funds, a sum of 50,000 yen, which will positively be returned to you within two months. Please give the money to the messenger. Of course we should visit you personally, but since we are so reduced to poverty that we are unable to do so and have asked the maid. Please, please help us.

From your former subordinates.

PS PLEASE GIVE US YOUR REPLY TO THE MESSENGER TODAY.

Ishii gave the letter, and many of the others that he received, to the Japanese police. Soon, U.S. Army Intelligence agents received microfilm copies of the letters.

By February 1947, enough new information had accumulated about Ishii's experimentation on humans that U.S. Army Intelligence headquarters in Tokyo decided that a third expedition from Camp Detrick was advisable in the near future. And so on April 15, 1947, Norbert H. Fell arrived in Japan.

Fell, a bacteriologist, was chief of the new Pilot Plant–Engineering Division at Camp Detrick. He held a Ph.D. from the University of Chicago, had been a commander in the U.S. Navy during the war, and was by far the most estimable of the Detrick interrogators to be sent to the Far East. He was a large man with a somewhat bulldog-like face and blunt personal manner, but he was well liked on post, where he was known as never being averse to having a drink.

Fell was not at all in the deferential mold of his predecessors, Sanders and Thompson. He was gruff, bullying, and not easily awed—besides which he could take notes in shorthand. If anyone could pry the remaining data out of Shiro Ishii, it would be Norbert Fell.

Fell spent his first few days in Tokyo reviewing the Army's files on the case. These convinced him that Ishii and company had depths yet unplumbed, that in fact the other investigators had only scratched the surface. But a new factor had emerged in the interim, one that might make all the difference in getting Ishii and the others to talk: the Soviets.

The Soviets were interested in Ishii, too, not so much for his value to the future progress of germ warfare but rather for his usefulness in war crimes trials. Both the Americans and the Russians were preparing to hold trials, and in the end more than 2,000 separate judicial proceedings would take place involving the Japanese. When the Soviets heard the tales of Ishii's medical experiments on POWs, they quite naturally wanted to include him in their roster of war criminals. Ishii, who was fully aware of their plans, was by this time "a thoroughly frightened individual," according to U.S. Army Intelligence agent Robert McQuail. McQuail and his friends, and now Norbert Fell along with them, reasoned that if they played upon Ishii's fears of being charged with war crimes, maybe they could induce him to reveal what he knew about the experiments on humans.

Frightened as he was, Ishii was not too scared to realize that he was still in a fairly strong bargaining position. He was the one with the data that everyone wanted, after all, information that he hadn't yet parted with despite having been grilled repeatedly by the American "expert," Arvo Thompson. Meanwhile he was not in prison; he was at his home in Tokyo living the life of Riley just as if nothing had ever happened. If the Americans were so hungry for the remaining data that they were coming back to Japan yet a third time, then they'd probably be willing to give him immunity for all of his past deeds, in return for which he'd tell them everything they wanted to know about the Japanese biological warfare experiments on humans and whatever else.

Before meeting with Ishii, Fell had a preliminary strategizing conference of his own with two of Ishii's former comrades: Tomosada Masuda, the Unit 731 commander whom Murray Sanders had interviewed at length, and Kanichiro Kamei, a Ping Fan alumnus who in the days since the end of the war had already become a successful businessman and politician. Both Masuda and Kamei were willing to help the Americans, but they were potential war criminals themselves, so they had already colluded with each other in an attempt to bargain with Fell. Kamei would act as the spokesman for the two, and he now related to Fell the gist of the deal.

"Masuda is eager to cooperate with you," said Kamei. "However, information on offensive developments in BW is extremely delicate, and Japanese formerly connected with this field are very loath to speak about it. Ishii was extremely unpopular with his subordinates, and one or more sent anonymous letters shortly after the surrender to SCAP, accusing Ishii of directing human experiments in BW and requesting that he be prosecuted as a war criminal. As a result, Japanese personnel were afraid to reveal information for fear of involving themselves or others. The interrogations conducted by Lt. Col. Sanders and Lt. Col. Thompson were too soon after the surrender. However, if the men who actually knew the detailed results of the experiments can be convinced that your investigation is from a purely scientific standpoint, I believe that you can get more information."

From a purely scientific standpoint, Fell knew, was a code phrase that meant, *Any information we give you cannot be used in war crimes prosecutions.*

"Masuda admitted to me that experiments were carried out on humans," Kamei continued. "The victims were Manchurian criminals who had been condemned to death. The personnel involved in carrying out these human experiments took a vow never to disclose information. However, I feel sure that if you handle your investigation from a scientific point of view, you can obtain detailed information."

Norbert Fell, it turned out, was fully content to handle his investigation from a scientific point of view.

On Thursday, May 8, 1947, Norbert Fell met Shiro Ishii. The encounter took place at Ishii's home, as had all of Arvo Thompson's interviews with Ishii the previous year.

Both men had had plenty of time to prepare themselves mentally and physically for the grand event, and both of them had rehearsed their parts quite well. This first conference, however, was to be no more than a short get-acquainted meeting, a chance for each to take the measure of the other and to lay a foundation for further talks.

It was also Fell's opportunity to see just how sick Shiro Ishii really was, and so Fell arrived at Ishii's house with an Army physician by the name of Captain Penton, plus Robert McQuail of Army Intelligence and Taro Yoshihashi, a U.S. Army interpreter.

Shiro Ishii, dressed in a kimono, was sitting up in bed; beside him was his own personal physician. Ishii did indeed appear to be in poor health, at least at first.

Fell now explained the ground rules to Ishii. He was here for "technical and scientific information," he said, "and not war crimes."

He was familiar with Ishii's previous testimony as given to Arvo Thompson, Fell said, and he knew what information he had concealed, especially concerning the human experiments and the use of biological warfare agents against the Chinese. Fell now wanted the full truth about all of this from Ishii. However, Ishii must not give any of the same information to the Russians.

Then it was Ishii's turn to speak.

"I will not reveal information to the Russians," he began. "I am responsible for all that went on at Ping Fan. I am willing to shoulder all responsibility. Neither my superiors or subordinates had anything to do with issuing instructions for experiments. If you ask me specific questions, I can tell you general results.

"I am wholly responsible for Ping Fan," Ishii repeated. "I do not want to see any of my subordinates and superiors get in trouble for what occurred. If you will give me documentary immunity for myself, superiors, and subordinates, I can get all the information for you."

Fell was prepared for this, but he was not prepared for the next words out of Ishii's mouth, which were: "I would like to be hired by the U.S. as a BW expert."

Hired by the U.S. as a BW expert?

"In the preparation for the war with Russia," said Ishii, "I can give you the advantage of my twenty years' research and experience. I have given a great deal of thought to tactical problems in the use and defense against BW. I have made studies on the best agents to be employed in various regions and in cold climates. I can write volumes about BW, including the little-thought-of strategic and tactical employment.

"With regard to anthrax," he continued, "I considered it the best agent because it could be produced in quantity, was resistant, retained its virulence, and was eighty to ninety percent fatal. The best epidemic disease I considered to be plague. The best vector disease I considered to be epidemic encephalitis."

Ishii went on like this for the next two hours.

Fell, at the end of it, told him that he had reached no final decision about "documentary immunity." He did, however, advise Ishii that the Soviets would be coming to question him soon. He also suggested that Ishii use "ill health" as an excuse to refuse interrogation by the Russians. In fact, it was for exactly this purpose that Fell had brought Captain Penton, the Army physician, along with him.

Penton now examined Ishii and pronounced him to be in good health. Thereupon Fell, McQuail, and Yoshihashi, the interpreter, coached Ishii as to how to feign and play up his illness whenever the Russians came calling. They told him to be vague and evasive in his responses, and to moan convincingly at appropriate points—as if Shiro Ishii needed help with any of this.

Then, their first bit of business concluded, and the meeting having gone to the satisfaction of all concerned, the four Americans left for the day.

When Norbert Fell got back to Detrick in June 1947, he, too, wrote up a report describing his accomplishments with the Japanese germ warfare hierarchy. "It was finally possible to get the key Japanese medical men who had been connected with B.W. to agree to reveal the entire story," he wrote.

His talks with General Ishii, "the dominant figure in the B.W. program," were particularly successful, he said. Ishii, according to Fell, was currently "writing a treatise on the whole subject. This work will include his ideas about the strategical and tactical use of B.W. weapons, how these weapons should be used in various geographical areas (particularly in cold climates), and a full description of his 'ABEDO' theory about biological warfare. This treatise will represent a broad outline of General Ishii's 20 years' experience in the B.W. field and will be available about 15 July."

There was more. Fell had already received a bunch of reports from some of the other Japanese biological warfare scientists, including one describing the results of nine years' worth of experiments on crop destruction, and another on the free balloon project in which "it was admitted that considerable attention had been given to using the balloons for dissemination of B.W. agents."

But the most important document of all was described as "a 60-page report in English on B.W. activities directed against man." This document included "full details and diagrams" of the Japanese experiments on human beings, starting with their use of anthrax bombs against human subjects.

"The human subjects were used in exactly the same manner as other experimental animals," wrote Fell. "In most cases the human subjects were tied to stakes and protected with helmets and body armor. The bombs of various types were exploded either statically, or with time fuses after being dropped from aircraft.

"The Japanese were not satisfied with the field trials with anthrax. However, in one trial with 13 subjects, 8 were killed in a result of wounds from the bombs, and 4 were infected by bomb fragments (3 of these 4 subjects died). In another trial with a more efficient bomb ('UJI') 6 of 20 subjects developed definite bacteremia, and 4 of those were considered to have been infected by the respiratory route; all four of those latter subjects died. However, those four subjects were only 25 meters from the nearest of the nine bombs that were exploded in a volley."

In addition to their work with bombs, the Japanese had also experimented with direct aerosol sprays of anthrax out of a flit gun. "In a typical experiment, four human subjects were placed in a glass room 10 m in size, and 300 cc of a 1 mgm/cc suspension were introduced using an ordinary disinfectant sprayer. No particle size determinations were made, but two of the four subjects developed skin lesions which eventually resulted in a generalized anthrax."

They had also performed plague experiments on humans, using the customary route of infection, fleas.

"Methods were developed for producing many kilograms of normal fleas (one gram @ 3,000 fleas), and for infecting them on a production basis. This flea work is described in great detail and represents an excellent study."

One part of this "excellent study" consisted of placing people in a flea-infested room, where "it was found that one flea bite per person usually caused infection. It was also found that if subjects moved freely around a room containing a concentration of 20 fleas per square meter, 6 of 10 subjects became infected and of those 4 died."

A second part of the "excellent study" consisted of trials with porcelain Uji bombs stuffed with fleas instead of the more usual anthrax slurry. "The fleas were mixed with sand before being filled into the bomb. About 80 percent of the fleas survived the explosion which was carried out in a 10-meter square chamber with 20 subjects. Eight of the 10 subjects received flea bites and became infected and 6 of the 8 died."

Gruesome as all of it was, Fell reported that there was yet much more to come, including a large selection of biological slides and specimens. Those samples were the fruit of many years' work and, as Arvo Thompson had correctly suspected, not all of the slides, samples, and records had been destroyed when the Japanese demolished the Ping Fan institute.

"It was disclosed that there were available approximately 8,000 slides representing pathological sections derived from more than 200 human cases of disease caused by various B.W. agents. These had been concealed in temples and buried in the mountains of southern Japan. The pathologist who performed or directed all of this work is engaged at the present time in recovering this material, photomicrographing the slides, and preparing a complete report in English, with descriptions of the slides, laboratory protocol, and case histories."

9

a t the end of the war, Ira Baldwin resigned his post as the scientific director at Camp Detrick and returned to the University of Wisconsin, where he took up residence as dean of the graduate school. He was not yet out of the bacteriological warfare business, however. He discovered, in fact, that immersion in the profession was all too much like being in the Mafia: once you were in, you were in for good. It was a highly specialized calling, and expertise in the field was a rare commodity much coveted by the military. So the U.S. Army couldn't let go of its grip quite yet.

In 1946, Baldwin and three other blue-blooded American germ warriors—George W. Merck, E. B. Fred, and William B. Sarles—wrote a paper called "Implications of Biological Warfare," a thoughtful reflection on what they'd collectively learned from their years in the trade. The moral of the story was that germ warfare was a technology that almost any nation could develop successfully. You didn't need anything on the order of the Manhattan Project to pursue it; indeed, the entire wartime budget for the Army's biological warfare program would have carried the Manhattan Project for all of a day or so. The raw materials of the enterprise—microbes—were out there in nature for the taking, and anyone with modest lab skills in biology could produce huge volumes of hot agents quickly and cheaply, and without any exotic hardware or ingredients. The most worrisome aspect of the whole business was that it could be done in secret, meaning that a given country's germ weapons program would be nearly impossible to detect, much less control or restrain.

But if it was easy to create the weapons in hiding, it was even easier to use them surreptitiously, for neither bombs nor explosions were required. Pathogens could be spread silently and invisibly from spray nozzles, and even the tiniest quantities could incapacitate or kill large numbers of people.

This meant that biological weapons were especially suited to "covert and clandestine operations"—a phrase that in short order became the new buzzword. To assess the magnitude of this threat, the National Military Establishment set up a new Committee on Biological Warfare, appointed Ira L. Baldwin as chairman, and charged the group with the task of estimating just how vulnerable the country was to covert and clandestine biological attack. In October 1948, Baldwin gave his answer in a "Report on Special BW Operations," a document that would take Camp Detrick in some new and bizarre directions.

"The United States," he wrote, "is particularly vulnerable to this type of attack. It is believed generally that espionage agents of foreign countries which are potential enemies of the United States are present already in this country. There appears to be no great barrier to prevent additional espionage agents from becoming established here and there is no control exercised over the movements of people within the United States." All this being true, it followed that "the subversive use of biological agents by a potential enemy presents a grave danger to the United States."

As to exactly how grave that danger was, there was no way of telling other than by means of experiment. You had to go out into the world and do the job, make an attempt at covertly exposing large masses of people to biological agents or trace chemicals. So Ira Baldwin now proposed that the country set up a program to spread noninfectious bacteria over the nation's cities, suburbs, and farmlands. Specifically, he proposed that undercover teams go out into the heartland and "test ventilating systems, subway systems, and water supply systems with innocuous organisms to determine quantitatively the extent to which subversive dissemination of pathogenic biological agents is possible."

Ira Baldwin, in other words, was now advocating that the United States undertake precisely the type of covert biological warfare trials that the Germans were alleged to have performed in the London Underground and the Paris Metro fifteen years earlier.

Moreover, that same cadre of special operations agents should also "determine quantitatively the extent to which contamination of intimately used objects such as stamps, envelopes, money, and cosmetics as a means of subversively disseminating biological agents is possible."

They should also try to infect large air masses. "Large air masses are constantly moving from the polar region over certain key areas of the United States," Baldwin said. "Possibly, these air masses could be utilized for the dispersion of BW agents."

Possibly, but who really knew? It was a scientific question, which meant that the only way to get a reliable answer was by actual trial. To banish uncertainty, you did the experiment. Now the entire country would serve as the lab.

On August 2, 1946, the Army had changed the name of the Chemical Warfare Service, the agency that supervised the biological warfare program, to something less warlike. It would henceforward be known as the Chemical Corps, a considerably more innocuous designation. In May 1949, in response to the recommendations contained in Ira Baldwin's "Special BW Operations" report, the Chemical Corps created an entirely new entity at Camp Detrick, a Special Operations Division—a sort of biological dirty tricks detachment.

No longer would the Detrick scientists focus exclusively on the recondite questions of pathogen selection, culture media, biological decay rates, and median lethal dosages for lab animals. Now they'd have to start thinking like spies. Which of course brought the CIA to Camp Detrick.

Camp Detrick's SO (Special Operations) Division was housed in Building 439, a long, low, yellow concrete-block lab structure, one that was identical in outside appearance to all the others on the base. Since the SO Division was to be a covert agency within the closed confines of an already highly secret military organization, the bulk of the personnel was recruited from other Detrick branches and divisions, and only the most trusted were invited to apply. Many who were accepted into the new outfit had been with the program from day one—Frank Olson, for example, who'd come to Detrick from the University of Wisconsin in 1943, and John Schwab, who was named the SO Division's chief. Schwab was one of the four people who'd grown the "three kilo dried X" for the British in the "Black Maria," an accomplishment that gave him an almost legendary status at Camp Detrick.

Once the division got up and running, the new department proceeded to fulfill the demands of Ira Baldwin's "Special BW Operations" report to the letter. The very first project on Baldwin's list had been to test ventilating systems, and this was in fact the first field operation that the SO Division carried out.

As for locale, what better ventilating system to test than the one in the Pentagon, the world's largest office building, where the U.S. Army itself had its headquarters? If this vast fortress was not safe from biological attack, what place was?

And so in the summer of 1949, John Schwab convinced the relevant Pentagon chiefs to permit a realistic test of the building's physical security. The agreement was that Schwab wouldn't tell them when or exactly how it would happen; the SO Division operatives would just come in and do their worst. In the event they were questioned by security guards, the men would carry official-looking letterheads from a fake air-quality testing firm, plus backup letters from the FBI explaining the true purpose of the operation.

The tests took place in August 1949, carried out by two- or three-man SO Division attack teams. The men were equipped with spray disseminators hidden in camera bags, suitcases, and so on, and air sampling devices that were likewise camouflaged. The suitcase samplers were the most troublesome items, since the battery-powered vacuum pumps drawing in the air made some noise. To take care of the problem, the attack teams left the units on the floor at strategic points with signs that said "Air pollution test." The security guards, who only wanted to help, often suggested the best setup locations along the miles of corridors.

Then the attack teams went around the building and calmly sprayed *Serratia marcescens* bacteria (SM) into the intake vents of the Pentagon's air-conditioning system. *Serratia* was the identical organism that the Germans had supposedly used in their raids on the subways of London and Paris. Back then it had been called *Micrococcus prodigiosus* (names changed fast in the rarefied world of biological taxonomy), but by any name the attraction of the microbe was that the organisms were bright red, making them exceptionally easy to track. The SO Division's spray devices were noiseless, the microbes being pushed out by compressed gas, and the men had no difficulty whatsoever in filling the U.S. Army's central command stronghold with billions of *Serratia* microbes. The Pentagon's rudimentary air filtration systems proved to be no use against the bacteria, and if the organisms had been anthrax spores instead of harmless simulants, they would have knocked out half of the country's top military ranks.

The nation's private citizens were just as vulnerable, however, as the SO Division's secret agents, helped by people from the Detrick M (Munitions) Division, demonstrated over the course of a three-week period in April 1950, when they sprayed whole cities with *Serratia marcescens* and *Bacillus globigii* (BG), a microbe which, because it was a spore former like *Bacillus anthracis*, made for a good anthrax simulant. The first attacks were made from the decks of the USS *Coral Sea* and USS *K. D. Bailey* anchored in the Atlantic off Hampton, Virginia. With the wind blowing in toward the shore, the microbes blew in from the sprayers and washed over the cities of Norfolk, Hampton, and Newport News.

That September, the Army and Navy repeated the experiment on the west coast of the United States, about two miles off the coast of San Francisco. Ships steamed up and down the coast spewing out three-mile-long lines of SM and BG, plus, for good measure, separate clouds of fluorescent particles (FP). The particles were composed of zinc cadmium sulfide, a chemical compound that glowed in the ultraviolet. They filtered down onto city streets and sidewalks, where after dark and under an ultraviolet light the scattered particles looked like stars on the ground.

The California tests were entirely successful, and traces of SM, BG, and FP were found as far away as twenty-three miles inland. "Nearly every one of the 800,000 people in San Francisco exposed to the cloud at normal breathing rate (ten liters per minute) inhaled 5000 or more fluorescent particles," said the official Camp Detrick report on the project. "The biological clouds released simultaneously with the inert aerosol covered very nearly the same area and presented similar dosage patterns."

Still, neither the San Francisco trials nor the earlier tests off the Virginia coast had addressed the question of the "large air masses" that Ira Baldwin had spoken about, the vast chunks of atmosphere that he said were "constantly moving from the polar region over certain key areas of the United States." To make a proper study of the phenomenon, the Army would have to do a large-scale dissemination by means of high-flying aircraft. In the late 1950s, accordingly, crews from Detrick bombed large stretches of the country with fluorescent particles in "Operation Large Area Coverage" (LAC).

"The first test took place on 2 December 1957," said an official Army Chemical Corps report. "A C-119 'Flying Boxcar' loaned to the Corps by the Air Force, flew along a path leading from South Dakota to International

Falls, Minnesota, dispersing fluorescent particles of zinc cadmium sulfide into the air. A large mass of cold air moving down from Canada carried particles along. Meteorologists expected the air mass to continue south across the United States, but instead it turned and went northeast, carrying the bulk of the material into Canada. The test was incomplete, but it was partially successful since some stations 1200 miles away in New York State detected the particles."

Another plane flew from Toledo, Ohio, to Abilene, Texas, spewing out a thin stream of FP at the rate of forty pounds per minute. Another flew from Detroit, Michigan, to Springfield, Illinois, and then turned west toward Goodland, Kansas. Sampling stations on the ground reported particles on both sides of the flight path, "proving that random flight over a target area would disperse small particles widely."

"These tests proved the feasibility of covering large areas of the country with BW agents," the Army report stated. "Many scientists and officers believed this was possible, but LAC provided the first proof."

Even this, however, constituted but a small fraction of the nation's total vulnerability picture, and the Detrick scientists would end up performing more than 200 simulant trials in, around, or upon the United States, sparing no corner of the country.

But as much as those tests revealed about the movement of aerosol clouds over the terrain and the persistence of those clouds over time, the results were fatally compromised by reason of having been obtained with simulants, materials that were thought to be inherently noninfectious, at least in the small quantities that people were intended to be exposed to. The scientific fact remained that simulant testing told you nothing much about the effects of a cloud composed of genuinely virulent pathogenic organisms.

By the time members of Detrick's SO Division were slinking around in the Pentagon with cans full of innocuous *Serratia* germs, the British, ever the pioneers in germ warfare, had been working with the real thing in the open air for more than a year. Not in England, of course, but in the balmy Caribbean. And, as always, the Americans were there to learn from the masters.

In 1948, when there was still a British Empire and Britannia ruled the waves, or at least the waters of its various outposts around the world, Paul Fildes and David Henderson at Porton Down conceived the idea of doing the next batch of hot agent trials on the open ocean. The Gruinard tests, successful as they'd been at the time, had been performed under highly limiting circumstances. The test site was within clear view of prying eyes on the shore, which was only a mile away. The trials could take place only when the winds blew from a given direction; otherwise the agent would be sent flying all over the mainland. Worst of all, Gruinard Island was now contaminated with anthrax spores, making it unfit for use by man or animal for God only knew how long in the future.

The trackless seas, by contrast, held many advantages for a set of germ warfare trials. You could do the tests in complete isolation, away from infectable humans, and out of sight. There would be no land area to worry about contaminating: any microbes that weren't inhaled by the test subjects—which could be positioned on rafts, boats, barges, or whatever—would fall harmlessly into the ocean, which would dilute the microbes to the point of noninfectiousness while the combined action of salt and sunlight destroyed all remaining traces of the agent. And if you chose the site carefully and with due malice aforethought, the winds would blow from a predictable direction for many months at a time.

So it was that on November 5, 1948, the Porton Down biological warfare fleet set sail for the British colony of Antigua, one of the northernmost islands of the Leeward chain, in the tropics, where the trade winds were moderate and fairly constant from the same direction and so would be ideal for carrying hot agents over long lines of animal specimens.

The Porton Down naval force consisted of three ships, the *White Sands Bay*, the *Ben Lomond*, and the *Narvik*. The focal point of the operation was the HMS *Ben Lomond*, a 4,000-ton landing ship whose network of masts, booms, winches, pulleys, and lines gave it the outward appearance of a freighter. In fact it was a laboratory ship, a floating germ warfare research facility that carried stocks of offensive agents, spray nozzles, air sampling devices, and munitions, including both British and American four-pound biological bombs, plus an assortment of smaller explosives. Its newly installed labs held centrifuges, dissection tables, surgical instruments, autoclaves, and a full complement of chemical and biological reagents and lab glassware. The *Ben Lomond*, in addition, boasted a deep freeze, a cold room, a central decontamination area, and the item without which no germ warfare facility was ever complete, an incinerator for animal carcasses.

It carried, in short, everything but the experimental animals themselves; they were aboard the *Narvik*, a landing ship the same size as the *Ben*

Lomond. The *Narvik* had been converted to a floating zoo, and on sailing day the boat had in its hold a total of 263 rhesus monkeys fresh from the jungles of India, plus a collection of guinea pigs and assorted other quadrupeds.

Accompanying the other two boats was the *White Sands Bay*, a Royal Navy frigate that acted as an escort vessel. The three ships made up Operation HARNESS, the first of five successive British germ warfare trials held on the high seas. In charge of every last one of them was John Dudley Morton.

Morton from Porton.

Morton was a thin and lanky specimen with a low, rich voice and a posh upper-class accent, the very image of bonhomie and politeness. He'd been educated at Cambridge where he'd taken an honors degree in organic chemistry with minors in botany, zoology, and meteorology. Upon graduation he came to Porton Down, where he specialized in the production and analysis of chemical smokescreens—optical barriers that an invading army could hide behind. Then, when it became clear that biological and not chemical warfare was the wave of the future, he taught himself microbiology, a subject he learned so well on his own that when David Henderson needed someone to plan and conduct his sea trials, he picked John Dudley Morton for the job.

On the *Ben Lomond*, Morton had his own private stateroom where he was attended by a servant named Taylor. One afternoon, five days at sea and 350 miles off the coast of Spain, Morton emerged onto the quarterdeck and was amazed to see, hove-to behind them, the HMS *Narvik* with its bow doors flapping open. The *Narvik* and the *Ben Lomond* had huge doors in the front, great metal clamshells that swung open laterally while a hinged ramp fell down into the water or onto a beach, for the discharge of amphibious craft, tanks, cargo, or personnel. The doors were not supposed to open spontaneously at sea, but there was the *Narvik* in just that condition, rolling and wallowing with stern to the wind, unable to move forward without taking on water.

The crew finally got some ropes around the doors and pulled them closed, and the ship was able to steam ahead safely at a speed of five knots. The flotilla detoured to Gibraltar for repairs.

The Porton fleet arrived in Antigua on November 30, after an ocean voyage of three weeks and some 4,600 miles. There on the dock waiting for them was David Henderson and a contingent of men from Camp Detrick.

The American government had maintained a military base on the island since World War II, an airport called Coolidge Field. The Detrick men— Arthur Gorelick, Al Webb, and a handful of Navy lab technicians—had been sent down there to set up a biological laboratory on a spit of land called Crabb's Peninsula across the harbor from the air base. The Americans were there to assist the British by growing hot agent and assaying cloud samples from the impinger flasks, in return for which they'd be allowed to get some practical experience at conducting open-air trials at sea.

Over the next three months, until the end of the following February, the Operation HARNESS team staged open-air germ raids on the waters adjacent to the three British islands of Antigua, Nevis, and St. Kitts. They disseminated a variety of materials, including white phosphorous smoke, biological simulants, and three hot agents: US (*Brucella abortus*, the cause of brucellosis), UL (*Pasteurella tularensis*, the cause of tularemia), and N (*Bacillus anthracis*).

The idea was to put the animals out on rubber dinghies, disperse the agent from a separate raft upwind of the animal floats, and then haul the exposed animals into the lab ship for the usual course of incubation-period monitoring followed by postmortems, incineration, and ritual scattering of the ashes. Meanwhile the sampling devices, which had been positioned next to the animals during the exposure, would be collected and sent back to the shore base where the Detrick men would culture the contents and tabulate the results.

The plan was plausible enough, but of course there were a few hitches. David Henderson contracted brucellosis. Animals fell into the water and drowned. Dinghies came loose from their moorings and floated off toward the horizon. Men in hot suits carrying heavy and fractious sheep nearly suffocated in the heat, got seasick, and sometimes vomited into their face masks.

Yet somehow it all worked. The key device was the trotline, a long line of rubber dinghies each connected to the next by a complex of ropes and electrical cables. Each little black raft held one or more experimental animals, boxed and immobilized as on Gruinard, together with an air sampler, a vacuum pump with which to draw air through it, and a battery to run the vacuum pump.

Working on the deck of the *Narvik* in hot suits consisting of white coveralls, boots, gloves, and face masks, the crew members boxed the animals and put them in the dinghies, which had been prepared the night before with fresh air samplers and charged batteries.

The ship's hinged front end now unfolded itself, the bow doors swung wide, the center ramp dropped into the waves, and the space-suited men on the foredeck were suddenly face-to-face with the open sea.

The dinghies had been arranged so that they could be pulled down the ramp and into the water, each dragging the next like a line of railroad cars. One of the ship's launches motored around to the front of the ship, took up the line to the first dinghy, and slipped away. The first raft splashed into the water and the rest of them followed in sequence. Soon the trotline was fully extended, some 200 feet long, twenty or thirty rubber boats floating along like a file of corks.

A second launch now towed the disseminator float into position. The float held the dispersal apparatus, which, depending on the experiment, was either a glass nebulizer known as a Collison sprayer, or a biological bomb.

Prior to the test, the *White Sands Bay* had made a reconnaissance run downwind to make sure the area was free of ships. The *Ben Lomond* and the *Narvik*, meanwhile, moved out of the path of the oncoming wind, so as to cause no turbulence in the smooth flow of air across the waves. By the time the trial began, the two mother ships and their launches were well to either side of the cloud's expected line of travel.

At time zero, the pathogen sprayed out from the source float and was carried along by the wind to the line of animals who breathed it in and retained some in their lungs. The rest of the cloud dispersed into the atmosphere or dropped into the water.

A winch on the *Ben Lomond* then pulled the trotline up the ramp and into the ship. The entire apparatus—boats, animal boxes, sampling devices, and batteries—was now hot with pathogen, and so the *Ben Lomond*'s decontamination team unloaded the animals and the samplers and washed down the dinghies, motors, batteries, and everything else with disinfectant. The animals went to a holding area on the dirty side of the ship. The trotlines went back to the *Narvik* for the next day's tests. The sampling devices were put on a launch for the shore base.

Chief of the shore base laboratory was Al Webb, a Detrick microbiologist with a doctorate from MIT. He hadn't been at Camp Detrick for much more than a year when Archie Gorelick, the chief veterinarian, came looking for volunteers for a mission to the tropics. The assignment, Gorelick told him, would involve scientists from three countries, England, Canada, and the

United States, working together doing aerobiology experiments on a secret island in the Caribbean that he was not at liberty to name.

A tropical island, warm weather, lovely beaches. Who could resist?

And so in November 1948, Webb, Gorelick, and a bunch of Detrick lab techs got aboard a four-engine MATS (Military Air Transport Service) DC-4 for a flight to Miami, then San Juan, and then, as it turned out, Antigua.

Coolidge Field was on a flat, isolated peninsula on the northeast side of the island. The airfield was also the site of their living quarters, which weren't much—just the usual enlisted men's barracks, Officers' Club, and Bachelor Officers' Quarters. Although he was a civilian, Webb was installed in the BOQ along with the rest of the Detrick crew.

The biological laboratory was across Parham Harbour on Crabb's Peninsula, a thin projection of grass, sand, and rock whose only land access was by means of a road that stopped at a guard post. The island's roads were so poor as to be barely passable, however, and so the men were taken across to the lab by boat.

The lab was located on a former U.S. Navy seaplane tending base, essentially a large concrete apron that ran down to the waterline. The Army now built a laboratory complex there solely for use in Operation HARNESS.

The lab building was a rectangular white wooden structure with slatted windows, shades, and mosquito screens. Inside was a full biological laboratory: lab benches, chairs, incubators, centrifuge, and the normal supply of glassware plus a few small animal cages. There was electricity and running water but none of the normal Detrick refinements such as negative air pressure and safety cabinets with exhaust hoods. Adjacent to the lab were two buildings full of animal rooms. They were the province of the Detrick vet Archie Gorelick and so the entire complex had come to be known as Archie's Place.

It was late in December when the first hot samples arrived from the *Ben Lomond*. The British had finally managed to lay down a cloud of *Brucella* over a trotline of monkeys. One dinghy had been lost in the operation, together with all of its sampling gear and animals. Still, the rest of the impingers and everything else had survived and they were now brought to the lab by launch.

Al Webb himself worked up the first sample. He took one of the impinger flasks and wiped the outside with Roccal solution, a disinfectant. Inside were about ten cubic centimeters of the sampling fluid.

He sucked up a quantity of the stuff using a mouth pipette, a thin glass tube with a piece of cotton wadding at the high end. Webb took a quantity of the liquid and transferred it out in equal portions to five small vials, then diluted the five samples in successive stages. The final step was to take a small amount of each dilution and transfer it to a petri dish filled with agar—a translucent jellylike substance—and incubate it overnight. Next morning, if *Brucella* bacteria had been present in the original sample, colonies of it would appear on the agar plates. From the size, number, and density of the colonies, he could estimate the concentration of bacteria to which the animals had been exposed.

When he opened the incubator the following morning, tiny colonies of *Brucella* had appeared on the plates. Webb filled out the report charts and sent them back on the next launch to the *Ben Lomond*.

Al Webb, Archie Gorelick, and the Detrick technicians spent the next three months in Antigua assaying impinger samples from the HARNESS trials and passing the microbes through various animal species. They also grew new hot agent, centrifuged it down to the proper concentration, and sent it back on the launch to John Dudley Morton on the *Ben Lomond*. They did all this with great attention to safety, and none of the Detrick crew ever got infected with the microbes they worked with on a daily basis.

Every now and then one or more of them would take the launch to the *Ben Lomond* for an observation and learning tour. And, as time progressed, various higher-ups from Camp Detrick, Aberdeen Proving Ground, and the Pentagon flew down to Antigua for an inspection trip. Charlie Phillips, Camp Detrick's decontamination expert, and Ken Calder, the Detrick meteorologist, paid a visit in December, as did General Alden Waitt, head of the whole U.S. Army Chemical Corps.

Henry Eigelsbach, Detrick's tularemia expert, came down in January, bringing with him a fresh supply of *Pasteurella tularensis*. Lord Stamp arrived from Porton Down, and at the end of February, Norbert Fell, who had interrogated Shiro Ishii in Tokyo two years earlier, showed up for the tour.

Operation HARNESS lasted until the end of February 1949. On the final day, the *Ben Lomond* anchored in the outer reach of Basseterre Bay, St. Kitts, set out the final trotline, detonated two 4-pound bombs full of UL, and refrigerated the impingers. The British sailed back to England, and the Americans flew home to Camp Detrick.

To John Dudley Morton, the mission was technically successful, although rather wasteful in terms of both personnel, of which there were hundreds, and lab animals, of which they'd gone through more than a thousand. And a small navy had had to sail across 4,600 miles of open ocean just to get to the test site.

The work, moreover, had been hazardous in the extreme. Getting the trot-lines in and out was a major operation: water sloshed in through the open doors of the landing ship, the ramp heaved up and down with the waves, the monkeys were semihysterical and hard to handle—they fought, spit, and bit—plus there was always the nightly nightmare of decontaminating the hot suits and the inflatable dinghies and everything else, and then hauling the entire trotline back over to the *Narvik*. As far as Morton from Porton was concerned, there had to be a better way.

There had been a final mission to Tokyo after Norbert Fell returned from Japan. This one was to get the slides.

Fell had reported that the Japanese had saved approximately 8,000 slides of pathological sections taken from some 200 human cases of dis-ease caused by a range of biological warfare agents. That, however, had proved to be a gross underestimate. There were in fact 15,000 slides from more than 500 patients, this out of an overall total of 850 human corpses upon which the Japanese scientists had performed autopsies. The American scientists wanted to get their hands on the slides, the test protocols, and the autopsy reports, and so on October 28, 1947, approximately four months after the departure of Norbert Fell for Camp Detrick, a new two-man team arrived in Japan.

Leader of the pair was Edwin V. Hill, M.D., chief of Basic Sciences at Detrick, a tall, graying man with glasses who also had a graduate chemistry degree from MIT. The other was Joseph Victor, M.D., a Detrick pathologist who had taught pathology at Columbia University's College of Physicians and Surgeons in New York. Although their primary objective was to collect the slides of human pathological material, Hill and Victor would also con-duct more than thirty new interrogations of Shiro Ishii and twenty other Japanese scientists who had worked in the germ warfare program at Ping Fan, Harbin, Mukden, or elsewhere.

One of the scientists, Masahiko Takahashi, had specialized in the aerosol delivery of infectious agents, a topic of surpassing interest at Camp Detrick. On November 20, Hill and Victor interviewed Takahashi, who described to them the chamber he used for human experimentation. It was octagonal shaped, he said, with a capacity of twenty-eight cubic meters. He would put a group of human subjects in the exposure chamber and then, using "an insect type sprayer similar to the flit gun," would introduce an aerosol of a given infectious agent at the rate of "approximately 1 cc of bacterial suspension per second." He experimented with sprays of plague, anthrax, typhus, smallpox, tuberculosis, and cholera, among other diseases, with varying rates of human mortality.

Another scientist, Shiro Kasahara, told Hill and Victor of his experiments with Songo fever, a recently discovered form of epidemic hemorrhagic fever that affected both humans and animals. He'd taken blood from humans suffering from the disease and injected it into horses; conversely, he'd taken blood from sick horses and injected it into humans. "Mortality of the natural disease in Japanese soldiers was 30 percent when the disease was first discovered," he told them. "However, mortality in experimental cases was 100 percent due to the procedure of sacrificing experimental subjects," which was to say, killing them.

The scientists took kidney, spleen, and liver sections from the sacrificed human cases, and mounted them on microscopic slides. Those slides were now in the hands of Tachio Ishikawa, the Unit 731 pathologist, who was photographing them for the Americans.

Midway through the process, Ishikawa had decided that he had so many more tissue sections than he had originally estimated that he'd need additional supplies to photograph them all. To help him along with his work, Lieutenant Colonel Robert McQuail of U.S. Army Intelligence had shipped him four boxes of photographic chemicals and other materials. Now, in November 1947, Hill and Victor, accompanied by their interpreter Taro Yoshihashi, visited Ishikawa in Kanazawa, a city on the west coast about two hundred miles from Tokyo, for a progress report.

Ishikawa was a pug-nosed and beefy man with close-cropped hair and short, thin eyebrows. At the time of his meeting with Hill and Victor, he was employed as a professor of pathology at Kanazawa University Medical School. When they arrived at Kanazawa, Ed Hill was disappointed that Ishikawa had not made more headway.

"The pathological material submitted to us in Kanazawa was in a completely disorganized condition," he later wrote. "It was necessary to arrange this material according to case number, tabulate the number of specimens, and inventory the specimens."

Looking at the mess of stuff in front of him, Victor commented to Ishikawa that he could tell which cases were controlled experiments by merely stacking the slides according to cases: the controlled experiments had many more slides, said Victor, to which Ishikawa smiled an assent.

Hill and Victor now collected the slides, the experimental protocols, and a number of autopsy reports, and brought them all back to Tokyo. The overall cache, Hill wrote later, "consists of specimens from approximately 500 human cases, only 400 of which have adequate material for study." He then listed in tabular form the total human cases of each disease as against the number of cases for which there was "adequate material for study."

HUMAN CASES

DISEASE	ADEQUATE MATERIAL	TOTAL
Anthrax	31	36
Botulism	0	2
Brucellosis	1	3
Carbon monoxide	0	1
Cholera	50	135
Dysentery	12	21
Glanders	20	22
Meningococcus	1	5
Mustard gas	16	16
Plague	42	180
Plague epidemic	64	66
Poisoning	0	2
Salmonella	11	14
Songo	52	101
Small Pox	2	4
Streptococcus	1	3
Suicide	11	30
Tetanus	14	32
Tick encephalitis	1	2
Tsutsugamushi	0	2
Tuberculosis	41	82
Typhoid	22	63
Typhus	9	26
Vaccination	2	2

There could be no doubt, at this point, that the principal Japanese germ warfare figures were war criminals of the greatest magnitude, on the order of Josef Mengele and the other Nazi doctors who had performed experiments of unimaginable cruelty on concentration camp prisoners during World War II. The Soviets, indeed, wanted to prosecute the Japanese Unit 731 hierarchy for their role in such crimes, yet the American biological scientists, in their rush to get the scientific data, showed no evidence of being held back by any moral, legal, or other constraints, and willingly promised immunity to Shiro Ishii and all the rest in order to get it.

This was motivated in part by lust for knowledge during the infancy of a new science in the aftermath of a major war, with signs on the horizon that the United States might yet have to defend itself against biological warfare attacks by the Soviet Union at some point in the near or distant future. Part of it of course may have been sheer paranoia. But from their own accounts, it's clear that another part of their motivation in gaining the data was humanitarian. Norbert Fell himself, in his report to the chief of the Chemical Corps, Alden Waitt, had explained that while the Americans were ahead of the Japanese in mass production of pathogens and munitions, the Japanese, precisely because of their use of human subjects, were ahead in research that could be of medical value to the Americans. The Japanese "data on human experiments, when we have correlated it with the data we and our allies have on animals, may prove invaluable," he said, "and the pathological studies and other information about human diseases may help materially in our attempts at developing really effective vaccines for anthrax, plague, and glanders."

Evidently, the American germ warfare researchers did not see it as part of their role as scientists to be concerned about the implications of gathering tainted data. Information was information, wherever it came from, and at all events the Americans had not *produced* the data, they were merely collecting it after the fact.

Even so, Detrick physician Edwin Hill did not have to go as far as he did when, at the close of his own report to the chief of the Chemical Corps, he ventured the opinion that the Faustian bargain they had struck had been, in the end, a tremendous financial deal. "Evidence gathered in this investigation has greatly supplemented and amplified previous aspects of this field," he wrote. "It represents data which have been obtained by Japanese scientists at the expenditure of many millions of dollars and years of work.

Information has accrued with respect to human susceptibility to these diseases as indicated by specific doses of bacteria. Such information could not be obtained in our own laboratories because of scruples attached to human experimentation. These data were secured with a total outlay of ¥250,000 to date, a mere pittance by comparison with the actual cost of the studies."

Further, Edwin Hill wanted no harm to befall Ishikawa, Ishii, or any of the other experimenters who produced the data that could be so useful to the Americans: "It is hoped that individuals who voluntarily contributed this information will be spared embarrassment because of it."

Spared embarrassment.

By any measure, the all-time low point at Camp Detrick.

10

b y the start of the 1950s, biological warfare had gone beyond the stage of a crash program aimed at developing a workable munition (which was all the British, Canadian, and American efforts had amounted to in the beginning) to a systematic effort directed toward creating the Ideal Biological Weapon. Such a device would disseminate the smallest viable product in the smallest possible amount over the target.

You wanted to use the smallest amount of live agent so as not to contaminate an area that was appreciably larger than the target itself. And you wanted to generate the smallest viable product because small-particle aerosols were far more efficient at infecting people than clouds composed of larger droplets. Particles that were one or two microns in diameter were better than those that were five or ten microns in size; the ultimate achievement, however, would be to encapsulate one live and undamaged pathogenic organism in each and every droplet of carrier liquid. One cell in one droplet—that was the platonic ideal.

The practical problem was how to make a weapon perform such a miracle. The answer, naturally, was by trial and error. You had to match the pathogen with the munition, employ the right amount of the proper explosive, optimize every controllable parameter, trying out all the assorted combinations to see what happened. That meant lots of tests—endless waves of bombs exploded over endless sets of live animals.

In the late 1940s, Herbert G. Tanner, chief of the M (Munitions) Division at Camp Detrick, had an idea for how to make any arbitrary number of such tests easily, at home base, without traveling repeatedly to places like Horn Island or Dugway, or going to the sea in ships as the British were then doing.

Tanner had once worked for the DuPont company in Delaware, where they stored propane, butane, and other gaseous or liquid chemicals in one-million-liter spherical tanks. The tanks were hollow forty-foot-wide balls propped up on posts that went around the circumference. Why not build such a thing at Detrick? If the walls were thick enough, you ought to be able to explode four-pound bombs inside it and do no harm to the interior. You could slide the animals in and out through an airlock, a true production-line operation.

Tanner and Harold Batchelor, another Detrick munitions man, drew up the specifications for this megasize, bombproof cloud chamber and submitted them up the ranks, and at length the Chemical Corps placed an order with the Chicago Bridge and Iron Works for a one-million-liter test sphere to be built in an open grassy area at Camp Detrick just a few hundred feet from where the "Black Maria" once stood.

The ball's components arrived in 1949, in about forty separate sections that lay on the ground upside down, like so many pieces of eggshell. The legs went up first, eight massive steel posts, each of them twenty-six inches in diameter. Then a construction crane lifted each curved piece into place one by one, and welders joined the parts together, making for airtight seals on every side. A rank of portholes appeared around the perimeter, then doors and hatchways, and miscellaneous other entrances and fittings. Soon a catwalk arose at the equator, making the globe resemble the planet Saturn. A second, smaller catwalk with outer railings took shape another story up, and then a small circular platform at the top. The final product looked very much as if a spaceship had landed at Camp Detrick and wobbled down on its eight supporting posts.

When it was finished in 1950 at a cost of $715,468, the test sphere was four stories high, had walls an inch and a quarter thick, and weighed 131 tons.

The British had the "piccolo," a little baby exposure tube into which the heads of three tiny mice could be stuffed all at once. The Americans had the "8-Ball," a one-million-liter minor planet.

The 8-Ball was so big that when it was filled with a cloud of vaporized biological agents, stratification layers appeared at different altitudes throughout the interior. The aerosol particles arranged themselves out in strata according to size, moisture levels, temperature, and so on, much as if there were a weather system forming up inside. The 8-Ball's internal environment was in fact a captive atmosphere, whose parameters were adjustable, within limits, at the will of the experimenter. The humidity could be varied between 30 and 100 percent, and the temperature between 55°F and 90°F—the desert to the tropics, or anything in between.

The final goal, however, was to set off biological bombs to create pathogen clouds of various densities, and for this purpose a "bombardier cabinet" was built at the south pole of the 8-Ball. This was a closed metal box with hatches at the ends and viewing holes and glove ports along the sides. The procedure was to place the bomb into the cabinet through the hatch, then, using the glove ports, load it with pathogen, and lift the device to the exact center of the 8-Ball by means of an internal hoist. After detonation, a system of internal fans would homogenize the cloud, the scientists would take samples of the aerosolized agent, and then they'd push the animals in for their timed exposure.

Because the technicians would be working just outside the 8-Ball's skin, the sphere had to be airtight, and so there were endless pressure tests with Freon, and then explosive tests with biological simulants—lots of bombs filled with *Serratia*—before any hot agents were actually used.

The first "hot shot," when it finally took place, was with *Pasteurella tularensis* ("UL"). Tularemia was the special province of Henry T. Eigelsbach, who cultured the organism in the tularemia lab building. On the morning of the shot, Edgar W. ("Bud") Larson, the aerobiology division chief, personally went over to Hank Eigelsbach's lab with an empty M114 in hand—M114 being the new name for the Mk I bomb that had originally started out in life as the British Type F four-pound biological bomb.

Eigelsbach had prepared a quantity of UL, and when Larson got there, an Erlenmeyer flask full of it was sitting under a negative-pressure hood. It was a white, milky-looking suspension, about a pint of it. Larson placed the bomb under the hood, removed the rubber diaphragm from the top end, and emptied the flask into the bomb cavity as if he were pouring himself a drink. He replaced the bomb's rubber stopper, then topped it off with a steel cap.

Eigelsbach handed him a towel soaked in phenol, a disinfectant, and Larson wrapped the bomb in the towel, removed it from under the hood, and carried it over to the 8-Ball building. Where, some minutes later, it exploded and sprayed out its contents. Light flashed through the portholes, there was a sudden loud report, and the technicians felt the blast with the soles of their feet. There were open vats of liquid disinfectant around the perimeter of the 8-Ball, and the shock wave had set up shallow wavelets on the surface.

The fans whirred inside, spreading aerosolized *tularensis* evenly through the test sphere. The next step was to add the animals. They were the province of Joe Jemski.

Joe Jemski's formal title at Camp Detrick was "Chief of the Test Sphere Branch," but he liked to refer to himself jokingly as "the man behind the 8-Ball." He'd been born in Massachusetts, grew up on the Lower East Side of New York, and then got a Ph.D. in medical microbiology from the University of Pennsylvania, a combination that gave him a rare accent indeed. Jemski had an inordinate fondness for animals. He loved being around them, he loved handling them and working with them.

Jemski was in charge of the subset of animals on the post who were awaiting their turn in the 8-Ball: mice, rabbits, guinea pigs, goats, sheep, burros, mules, and birds, plus a collection of primates including monkeys and chimpanzees. He once selected a monkey from out of the bunch, a *Macaca fasicularis* from the Philippines, and made it his personal pet, naming it "Joe," after himself. Every morning when Jemski came in, the first thing he did was to check up on all the various animals under his care. "Joe" was small enough to sit on Jemski's shoulder, so "Joe" was always perched there as Jemski made the rounds from cage to cage. The other monkeys absolutely hated him for this privilege and hissed and spit at him from their cages.

There was also a chimp called "Grandma," the temporary pet of an animal handler by the name of Gibson. "Grandma" was friendly and cooperative, and every now and then she took up a broom and, mimicking her handler, swept the floor. Joe Jemski thought this was quite charming and so he had developed a special fondness for "Grandma," too.

The other part of Jemski's job, however, included placing animals into the 8-Ball for their "aerosol challenge," as they called it in the trade. You did not refer to this as "gassing the animals." The animals were not being "gassed," they were being *exposed to a biological agent*, one that, true enough, was more often than not an infectious organism that would sicken or even kill them. But animal exposures were an unavoidable and necessary part of both the offensive and defensive branches of biological warfare research. To judge the effectiveness of a biological weapon, you had to fire it upon actual living subjects, and for obvious reasons you did not want those subjects to be human beings. And likewise to produce a vaccine or toxoid, you wanted to test it out first on animals before subjecting any human volunteers to the pathogen in question.

So when aerosol challenge time rolled around, Jemski or one or more of his subordinates would go over to the animal room, or the goat corrals, or the sheep pens, and bring back whatever was needed. The 8-Ball was designed so that it could expose virtually any animal, man or mouse, to virtually any pathogen. Not all the animals would be exposed in the same manner, however. The smaller mammals—mice, rabbits, guinea pigs, monkeys—would be conveyed through an airlock for a so-called whole-body exposure.

The airlock was called a "transfer neck" and consisted of a cylindrical chamber about two feet long and a foot in diameter. There were a number of transfer necks at different points around the equator, where all the exposures took place. The caged animals went in through the outer door of the transfer neck, and then the hatch was closed behind them. Once the cloud had been put up and was circulating smoothly inside the 8-Ball, the airlock's inner door was opened and the cage was pushed out into the darkness (the internal lights had been turned off to protect the agent) along a small metal track.

The animals hung there in space, twenty feet above the bottom.

They took a one-minute whiff of the gaseous envelope, a dismal mix of air, explosive gases, slurry liquid, and infectious organisms encased in their droplets, after which the animals were brought out again and transferred to their hermetically sealed, climate-controlled confinement cages. Then, like every other test subject in the biological warfare program, they either died from the agent or were sacrificed after the incubation period had passed, and then autopsied and incinerated.

Large animals—sheep, goats, and burros—were too big and bulky for a whole-body exposure in the 8-Ball. They were exposed by mask instead. The mask was made of clear plastic and conformed to the shape of the animal's head, and so it was long and pointy, like a horse's feedbag. The mask went over the beast's nose, straps went around the back of the head to hold it in place, and a rubber ring that ran around the edge of the mask was pulled closed to make an airtight seal.

The large animals did not go quietly, however. They drooled into their masks, they hyperventilated, they stomped, kicked, and bucked, in some cases narrowly missing their handlers.

Falling back on what animal psychology skills they possessed, the handlers proceeded to introduce the beasts to their fate on a gradual basis, conditioning them with "pre-training experiences" and dry runs. Well before the test date they brought the beast to the sphere for a brief look-around and get-acquainted tour of the place. They'd walk the animal from the pasture over to the 8-Ball building and take it up in the elevator to the equator level. They'd lead it around to the exposure cubicle, let it sniff around at things, and then put its face mask on. They did this again and again until the beast was more or less comfortable with the program and breathing normally—whereupon it was exposed to the agent.

At various places around the equator, pieces of clear plastic tubing ran straight through the wall of the 8-Ball and into the interior. The tube had a clamp at the end so that none of the atmosphere leaked out, but with a face mask attached to the tube, the clamp could be opened, putting the wearer in direct contact with the internal atmosphere, whatever it was, and by this means the animal ended up breathing its allotted dose of pathogen.

That left the chimps, who had their own special inhalation arrangements at the 8-Ball. In the early 1950s, Air Force veterinarians at Holloman Air Force Base, near Alamogordo, New Mexico, had started training chimpanzees for their heroic contributions to the American manned space flight program. They joked about running a "Chimp Academy," about "astrochimp training" and "chimponauts," but eventually, in 1961, one of the chimps, "Ham," actually made it into space. During training the animals were seated in a "chimp chair," a stainless steel straight-backed throne whose nylon bindings and other restraints gave the device a marked resemblance to the electric chair.

The Detrick researchers decided that the Air Force chimp chair was custom-made for the 8-Ball, and so the fabricators in the E (Engineering) Division got hold of the specifications and put one together. It worked like a charm. The researchers seated the animal in the chimp chair, strapped it in head and foot, and belted its arms to the metal armrests. They placed a face mask on the chimp's head, snugged it tight, then bolted the mask to the back of the device. At this point the animal could not budge. It could not do anything but sit there and inhale whatever came in through the plastic tube.

Which the Detrick chimpanzees did, one after the other, even including "Grandma," the aerobiology division's pet chimp. The day finally came when her number was up, and so "Grandma" was strapped into the chimp chair, snugged down, bolted back, and then doused with anthrax or *Brucella* or *tularensis* or whatever else they were serving that day, and then she, too, went on up to chimp heaven.

Finally it was even time for "Joe," Joe Jemski's personal pet monkey, to have his "aerosol challenge." Being a monkey and not a chimpanzee, "Joe" would not get to sit in the chimp chair; rather, he would go directly into the 8-Ball itself for a whole-body exposure.

Jemski was saving "Joe" for last, but eventually the time arrived, "Joe" took his turn in the chamber, and that was that.

The 8-Ball ran through a sizable population of Camp Detrick test animals. The anthrax trials alone consumed more than 2,000 rhesus monkeys.

Even to Joe Jemski, that was a lot of monkeys.

The 8-Ball went a long way toward creating the ideal biological bomb—at least the ideal *laboratory* bomb. But a device that worked well in the highly artificial and controlled surroundings of the Camp Detrick test sphere was not necessarily the same thing as one that performed well in the field. As the nation's military commanders would shortly learn, biological bombs were an entirely different order of beast from the ordinary high-explosive bombs that they had come to know and love over the years.

Conventional bombs could be made by the million, stored anywhere, and pulled off the shelf and used when needed. Until the moment they exploded, they were inert objects that would last unchanged for decades, and perhaps even centuries. Biological bombs, on the other hand, contained living organisms, entities that were highly temporary and perishable. In order to do their work, those organisms had to be actively kept alive, meaning that certain environmental conditions had to be provided or else the steel shell containing them, despite all its external similarities to a conventional bomb, would be no more lethal than a flower pot. Biological bombs, in other words, were creatures of an hour.

There were exceptions: a bomb filled with anthrax spores, for example, would have a shelf life measured in decades. But the first biological bomb to be standardized by the U.S. military was not an anthrax bomb. The anthrax bacillus, after all, was a lethal organism, meaning that a bomb filled with it was an essentially redundant munition: it would do only what conventional bombs already did—kill people—albeit more slowly. For use in future wars, the U.S. Air Force wanted the option of drawing upon a "balanced arsenal," one that held a diversity of weapons having different operational characteristics and a variety of final results. They wanted, in particular, a weapon that would incapacitate rather than kill. And so in 1949, when the U.S. Army Chemical Corps finally decided on the first biological agent that it would formally standardize and make available for use as a weapon, they chose not a lethal organism but rather the incapacitating bacterium *Brucella suis.*

Brucella suis ("US") caused brucellosis, an animal disease that affected mainly cattle, sheep, and goats, among whom it caused spontaneous abortions. Humans contracted brucellosis by eating meat or milk products from infected animals, by direct contact with them, or by inhaling aerosols containing the microbe.

As a vehicle for a disease to be produced intentionally by means of biological bombing, *Brucella* was viewed as a kindly, mild, and humane microbe. The organism normally only made people sick; it rarely killed them. Since a human population could be disarmed just as well by sickness as by death, incapacitating agents were obviously preferable to lethal ones on humanitarian and ethical grounds. Also, brucellosis was not communicable from person to person. The initial bacterial cloud would infect a discrete and limited community and would go no further; there would be no secondary round of infection, no chain-reaction epidemic of the disease.

Despite these "humane" features, brucellosis victims nevertheless suffered high fever, sweats, fatigue, loss of appetite, muscle aches and pains, plus depression, headache, and irritability, along with local infections of the bones, joints, and the genitourinary tract. The symptoms often lasted for three to six months, and sometimes even for a year or longer. Also, the illness arrived and departed in waves, undulating like a sine curve, for which attribute brucellosis was also known as undulant fever.

From a mass-production standpoint, *Brucella* bacteria were relatively easy to grow, especially *Brucella suis*, which was a strain known to be highly pathogenic for humans even though it normally affected pigs (*suis* meant swine). Early studies at Camp Detrick had shown that the organism grew like wildfire, with a two-hour doubling time, in the presence of adequate aeration.

From a dissemination standpoint, the agent worked well enough in field trials—other than for the puzzling episode at the Suffield Experimental Station in Canada when not a single one of the thirty-five baby hogs set out for one test had been infected by the pathogen cloud. Still, that had been in the early days, 1946, an anomaly in an otherwise acceptable track record, and the scientists had often gotten excellent *Brucella* infection rates among the test animals.

From the medical, production, and dissemination standpoints, then, *Brucella* made a fine weapon. It was only from the *operational* standpoint that *Brucella suis* was a nightmare. Bombs filled with the agent had a finite shelf life of a few months: exceed it, and the device was a dud. Store the bombs at too high a temperature, in the midsummer sun, for example, and the shelf life would be proportionately reduced. The life span could be lengthened by refrigeration, but no one in their right mind wanted to store a million or more *Brucella* bombs in meat lockers. This, however, meant that such weapons could not be stockpiled at all, which was a highly novel concept for the U.S. Air Force.

Fighting a war with *Brucella* bombs, they soon learned, would require some highly offbeat military logistics. You would have to build mammoth bacterial production facilities and keep them on indefinite standby, ready to swing into mass production on short notice. Separately, you'd have to maintain an inventory of empty bombs that could be filled with the agent as soon as it was produced. Then, when you actually needed a supply of live bombs, you'd grow the bacteria, fill the empty bomb casings with it, gang the filled bombs together into clusters, airlift the clusters to the combat zone, load them onto bombers, and finally drop them on the enemy. It was a classic Rube Goldberg operation fraught with endless possibilities for things to go wrong, including accidents and sabotage, at every successive stage of the game.

But if the Air Force wanted an incapacitating biological device in its inventory, *Brucella* was what the Army Chemical Corps was offering to fill it with. In 1951, therefore, the Air Force formally standardized the *Brucella suis* biological bomb, which thus became the official American biological weapon.

Naturally, it would have to be field tested. In 1950 and again in 1951, the Army and Air Force had dropped some experimental *Brucella*-laden bombs out of B-29s at Dugway Proving Ground in Utah. The results were satisfactory as far as they went, but they only established that the device itself would work when finally dropped, not the antecedent bacterial growth, filling, clustering, and transport operation. The entire long Byzantine scheme would have to be proven out and demonstrated in a realistic field trial. This was scheduled for the summer of 1952.

The star of the show, the brucellosis agent itself, would be furnished by Camp Detrick, where yet another new pilot plant had risen on the grounds, Building 470, a seven-story redbrick windowless structure, constructed solely to support the Air Force requirement for a supply of biological weapons. The Detrick crews would also load the new-grown agent into 1,100 M114 four-pound bomblets. From that point on, however, the Detrick scientists were out of the picture.

The M114 bombs, in the Air Force plan, were to be dropped not individually but rather in clusters of 108 bomblets bound together in the Army's M26 cluster adapter, giving the final and finished device, the M33/*Brucella* cluster bomb.

The M33, which weighed 500 pounds when fully loaded, would be released from the aircraft at high altitude, its individual M114 bomblets would be ejected out of the canister before reaching the ground, and each bomblet would be exploded separately by its own detonator. Munitions experts had calculated that to properly infect one square mile with the requisite amount of biological agent, eight to sixteen such clusters, or 800 to 1,700 individual bomblets, would be required.

In the operational test to be conducted in the summer of 1952, the filled bombs would be put into clusters not at Detrick but at Edgewood Arsenal, some sixty miles away. This meant that some 1,100 M114 bomblets filled with *Brucella suis* suspension would have to be transported by truck over the Maryland highways from Detrick to Edgewood.

Still, the bomb drop itself would not take place at Edgewood but at Dugway Proving Ground, in Utah, about 2,000 miles away. For the test to be a realistic demonstration of the overall battle plan, however, it would not be permissible to fly the clusters directly from Edgewood to Dugway, the reason being that in an actual live war the clusters would be transported first to an intermediate staging point near the battle zone and only then loaded into bombers for dropping out over enemy territory. They'd go from Edgewood to Europe, for example, aboard transport planes, then be removed to temporary storage, and finally be transferred to the bombers that would drop them over Mother Russia or wherever.

So to simulate transport to an intermediate staging point, the clusters would be flown first to Florida, to Eglin Air Force Base. There the clusters would be unloaded from the transport planes, held overnight in refrigerated vans, and the next day loaded into the bomb bay of a B-50 biological bomber for an operational strike at the target. Just to be on the safe side while this bacterial freight was crisscrossing the United States, a separate technical escort, consisting of safety personnel, decontamination equipment, and supplies, would follow along in a C-124 cargo plane.

That was the sequence: Detrick, Edgewood, Eglin, Dugway—where, on the ground, a highly unusual target awaited them.

The bombing zone at Dugway for the operational suitability tests of the M33/*Brucella* munition was a level stretch of scrub desert in the shadow of Granite Peak. There, inscribed in the sand, was a circle 8,000 feet in diameter, the bull's-eye.

Superimposed on the circle, its four vertices extending slightly beyond it, was a square grid 6,000 feet per side. Inside the grid, distributed at 1,000-foot intervals in seven rows of seven columns throughout the square, were forty-nine identical plywood houses. Each house was a cube exactly fifteen feet ten inches on a side, and contained one door, eight windows, a wooden floor, and a ventilator in the roof. The doorways of adjacent houses in the same row faced in opposite directions. For the test, all the windows and doors were closed.

At the center of each house, on a wooden platform five feet high, were ten guinea pigs, boxed and in cages. The guinea pigs were of different strains and different sexes, and weighed between 350 and 600 grams apiece.

This was the enemy city.

Not all of the population was indoors, however. In front of each house was a trench ten feet long, two feet wide, and four feet deep. The trenches were arranged so that in four of the seven rows of houses, they ran parallel to, and were perfectly centered in front of, one side of the building. The trenches in the three remaining rows ran at right angles to the respective houses. In every case, no matter in what direction they were oriented, the center of the trench was ten feet from the side of the building. At the bottom of each trench were ten more boxed guinea pigs—part of the enemy city's outdoor population.

In addition, between each trench and the house it was in front of was a five-foot-high wooden platform. These outdoor platforms were identical in size, shape, and height to those inside the forty-nine houses, and each of them held ten further boxed guinea pigs. Finally, set out at 500-foot intervals throughout the city were a number of additional five-foot-high wooden platforms, each of them bearing their separate load of guinea pigs like the rest. At the time of the first operational test of the M33/*Brucella* munition, therefore, the total indoor and outdoor population of this enemy city on the desert plain would be 3,230 guinea pigs.

Then there was the lighting—not for the animals but for the bombing crew. The bombing runs would be staged at night, the reason being that *Brucella* was highly sensitive to sunlight. That, indeed, had been another reason why the Army had selected it as their first standardized incapacitating agent: the infectiousness would be localized around the target area itself, for the next day's sunlight would burn off and neutralize the remainder.

And so to guide the B-50 biological bomber to the target as it flew from Eglin Air Force Base at an altitude of 20,000 feet, there were two concentric circles of white lights stationed within the array, an inner ring of twelve 100-watt light bulbs at a radius of 1,500 feet from the center and another, bigger circle at 3,000 feet, all of them powered by two generators at opposite ends of the grid. In addition, as an aid to orientation, there was a straight line of lights strung along the grid's north-south axis. In the exact center of the bull's-eye, at ground zero, there was a single bright 1,500-watt lightbulb. From the air at night, the enemy city would show up as a white snowflake spread out on the black earth.

After the bombs exploded, the pathogen cloud would move across the city and then proceed off in the direction of the ground-level winds at the time of the test. Those winds would be from the south, and so for the purpose of sampling the cloud's density as it moved over the earth there were three "downwind sampling arrays" set out in arcs positioned two, three, and four miles north of the target center. The sampling arrays consisted of petri dishes filled with an agar medium. There were additional petri dishes on each animal platform within the enemy city—a combined total of 482 petri dishes open to the night sky. Including both the petri dishes and the animals, there were 3,712 individual sampling points for this one test.

On August 9, 1952, at 1:01 A.M., the B-50 biological bomber from Eglin Air Force Base arrived over the test grid, opened its bomb bay doors, and released its clusters on the target.

That was the state of American biological warfare combat-readiness in the summer of 1952—a mock attack with an incapacitating bacterium against 3,000 boxed guinea pigs. At that precise point in history, the Korean War had been going on for some two years. The war had begun on June 25, 1950, when 38,000 North Korean troops backed by fifty Soviet tanks advanced over the 38th parallel and headed toward Seoul, capital of South Korea, their object being to reunite the country under the rule of the North.

On June 26, the UN Security Council formally branded North Korea as the aggressor and called upon its member nations to join together and repel the assault. Fifty-three UN member nations supported the Security Council resolution, twenty-nine of them sent aid in one form or another, and sixteen would ultimately wage actual battle against North Korea: the United States, England, Australia, New Zealand, Canada, Turkey, Greece, France, Belgium, Luxembourg, the Netherlands, Thailand, the Philippines, Colombia, Ethiopia, and the Union of South Africa. In addition, India, Sweden, Norway, Denmark, and Italy supplied medical assistance.

So a bare four years after the end of World War II, and seven years after it had entered the biological warfare business, the United States was once again engaged in a major new shooting war. Its germ warfare technology had developed considerably in the interim. Pilot plants had gone up, agents produced, munitions tested. The Army had spent millions of dollars and countless man-hours and used hundreds of thousands of live animals on the research, testing, and weaponization of a wide variety of lethal and incapacitating pathogens, but not once during all that time had it ever used the stuff on the battlefield.

Whereas, by contrast, the instant it had a working atomic bomb, the Americans had used it not once but twice.

And so of course you had to wonder. Wasn't the Korean conflict opportunity knocking? And if it was, would the American biological high command be able to resist?

Hardly a year into the war, the North Korean government charged that the United States was waging biological warfare against them. On May 8,

1951, North Korea's minister of foreign affairs, Pak Hen Yen, sent a cable to the president of the UN Security Council saying that the United States had attacked Pyongyang and the surrounding areas with weapons carrying the smallpox virus. The attacks occurred between December 1950 and January 1951, he said, during which time some 3,500 cases of smallpox had broken out in the area, with 350 deaths.

The charge was answered by the United Nations Commander, General Matthew B. Ridgway, who had taken over from General MacArthur in April 1951. In May, Ridgway called the charges "deliberate lies." Neither the American forces nor any of the forces under his command had used biological weapons against anyone at any time. Moreover, smallpox was endemic in the affected areas and these were natural outbreaks that the North Koreans were portraying as biological attacks for propaganda purposes.

After those outbursts, nothing more was heard about biological warfare for the next several months. In February 1952, however, North Korea's Pak Hen Yen was back with new claims. "The forces of the American imperialist invaders again used bacteriological weapons early this year for mass annihilation of the people," he said. "The American imperialist invaders, since January 23 this year, have been systematically scattering large quantities of bacteria-carrying insects by aircraft in order to disseminate infectious diseases over our front line positions and rear."

Bacteriological tests showed, he claimed, that the insects were infected with plague, cholera, and other diseases.

Shortly afterward, in March, Chou En-lai, minister for foreign affairs of the People's Republic of China, claimed that the United States was also bombing China with bacterial weapons. The Americans were using not only insects but a fantastic assortment of bombs and vectors in an attempt to spread smallpox, anthrax, plague, meningitis, encephalitis, and cholera, along with fowl septicemia and four different diseases of plants.

These charges were soon accompanied by details which, if authentic, would indeed appear to show that American bombers had flown over the area and dropped the biological kitchen sink upon the Far East. There were eyewitnesses, photographs, lab results, and artifacts.

On March 12, 1952, according to local public health authorities, residents of K'uan Tien, a town near the Yalu River in the Liaotung Province of China, watched as eight American F-86 jet fighters flew overhead and one of them released a bright cylindrical canister. This was in broad daylight, shortly after noon, and the witnesses set out to find the object. They failed.

What they found, instead, were swarms of anthomyiid flies and masses of spiders.

Everything about these insects was anomalous: they appeared in habitats that were wrong for the species, at the wrong time of the season (some of the insects had been found on snowbanks), and in concentrations, combinations, and distribution patterns that were irregular in the extreme.

A separate group of investigators found quantities of fowl feathers scattered in the vicinity. When the flies, spiders, and feathers were examined in the laboratory, all of them showed traces of the anthrax microbe, *Bacillus anthracis*.

Nine days after the incident, a search party located a small depression in the ground, one that looked as if it had been made by a bomb. Inside the crater, and on the surface of the snow-covered cornfield that surrounded it, were the remains of an object that apparently had shattered upon impact. Spread out in a circle were some 200 pieces of a thin white chalky material, plus a metal rod and a cap-shaped steel plate. Later researchers would characterize the object as an "artificial eggshell."

Even before they'd found the eggshell, members of the search team had fallen ill with a respiratory infection. After a short course of the disease, four of them were dead of pulmonary anthrax. That, anyway, was the story. The incident was one of many insect scatterings alleged to have taken place along the border between China and North Korea that spring.

And then there was the episode of the voles.

On April 4, 1952, around midnight, residents of Min-chung, a village in northeast China, heard the sound of an aircraft flying over the rooftops. Members of the Chinese Air Observer Corps identified the plane as an American F-82 night fighter. Next morning, the village of Min-chung was crawling with voles. There were voles in the streets, voles in the fields, and voles on the frozen surfaces of ponds and wells. Dead voles were found on the rooftops and inside the houses on people's beds.

Altogether, a total of 717 voles were found in Min-chung and nearby settlements. Many of those that were still alive had broken legs, as if they'd been dropped from a height. All the rodents were of the same biological variety, a species that none of the villagers had ever before seen.

An epidemic prevention team now arrived in Min-chung, burned the dead voles, and buried the ashes. They tested a lone survivor for *Pasteurella pestis*, the plague bacillus, and discovered that it contained traces of the bacterium. However, due to their public health efforts, and also because no fleas had been found with the voles in Min-chung, no plague epidemic broke out in the village. Still, the event bore a disturbing similarity to the Japanese plague-bombing of Changteh in 1941.

And then, finally, there was the tale of the clams.

On the night of May 16, 1952, residents of Dai-Dong village, a community in rural North Korea, heard an airplane circling above them for about an hour. Early the next morning a young woman picking herbs came upon an unusual straw box on a hillside. She opened the package and found that it was full of clams. She took them home and she and her husband made a meal of the shellfish, eating them raw.

Soon the two were taken ill, as if the clams had been bad or poisoned, and by the evening of the following day both she and her husband were dead. Medical examination showed that the cause of death was cholera.

Meanwhile, four more straw packages full of clams had turned up on local hillsides. Some of the mollusks had broken shells, as if they'd fallen to the ground from a height. These clams were taken to a medical laboratory for analysis, where it was found that all of them were infected with *Vibrio cholerae*, the cholera bacterium.

Anthrax-covered insects, plague-ridden voles, cholera-infected clams. American planes flew over and dropped objects, diseases broke out, and people died. As grist for a biological warfare charge, the evidence was impressive and altogether incriminating. Almost too good to be true.

11

b ut the case of the insects, the episode of the voles, and the tale of the
clams were not even the tip of the iceberg. On March 8, 1952, Chou En-
lai claimed that in the six-day period between February 29 and March
5, U.S. biological bombers had made a total of 448 sorties into China, dropping flies, fleas, mosquitoes, and an assortment of other insects on the civilian population of his country. He later expanded the list to include lice,
locusts, mites, springtails, crickets, beetles, ants, caterpillars, sandflies,
butterflies, and bees.

And those were only the insects. The American bombers, according to
various statements issued by the Chinese Communists and North Koreans,
were scattering an absolutely incredible assortment of disease carriers
upon their lands: paper envelopes, straw, grain, cornstalks, bean stalks,
medical goods, cloth, candy, dead branches, leaves, manure clumps, crys-
tals, yellow powder, contaminated meat, earthworms, frogs, birds, gray
mice, rabbits, foxes, dead pigs, toilet paper, and infected pancakes.
Practically the only item the United States was not charged with dropping
was the single weapon that it would shortly standardize for battlefield use,
the M33/*Brucella* combination.

So serious were these allegations that the International Association of
Democratic Lawyers, a left-leaning advocacy group based in New York,

decided that it had to investigate this campaign of biological terror evidently being waged by the U.S. government against their heroic and peace-loving comrades in the Far East. So the democratic lawyers put together a special delegation of experts and sent it off to personally behold the scene of the crime.

The group, composed of attorneys, judges, and law professors from eight different countries including Poland and Communist China (but none from the United States) arrived in Korea in March 1952 and started taking evidence from eyewitnesses. The eight members, they later reported, were "impressed by the clarity and obvious sincerity and veracity of the many simple peasants and others who gave evidence as to the facts." The members also met with North Korean public health workers and examined the various official documents they provided, and then moved to China where they repeated the whole process.

The junket ended three weeks later in Peking, at which time and place the lawyers issued not one but two formal reports condemning the behavior of the American military in Korea and China. The first, "Report on U.S. Crimes in Korea," issued on March 31, 1952, claimed that the U.S. Army was conducting not only biological warfare in North Korea but chemical warfare as well. American B-29s, it said, had dropped chemical gas bombs on Nampo; U.S. jets had spread noxious chemicals over Poong-Po Ri; planes had flown over Hwanghai Province and yellowish-green clouds suddenly rose overhead, after which four people died, tree leaves dropped to the ground, grain crops withered, and brass objects turned black.

The second document, "Report on the Use of Bacterial Weapons in Chinese Territory by the Armed Forces of the United States," released in Peking on April 2, 1952, essentially repeated the official Chinese account of the American biological war crimes in their country.

The impact of the two reports was somewhat blunted by a front-page story in the *New York Times* the following day. On March 15, 1952, when the democratic lawyers were about midway into their fact-finding tour, the *Peiping People's Daily* had published nine photographs of objects "dropped by United States invaders" along with captions identifying the items as "tiny black insect," several "poisonous insects," and flies that very suspiciously "could crawl but not fly." In addition, there were images of "meningitis double globular bacteria," "consecutive-globular bacteria," and "bluish impurities," among other things.

The *Times* sent the photographs together with a translation of the captions to two U.S. experts, Dr. C. H. Curran, chief curator of insects and spiders at the American Museum of Natural History, and Dr. René Dubos, a bacteriologist at the Rockefeller Institute for Medical Research in New York. In a page-one story published on April 3 under the headline "Reds' Photographs on Germ Warfare Exposed As Fakes," the *Times* reproduced the photographs and captions together with the comments of the two experts.

The "tiny black insect" was especially noteworthy. It looked like a horror film monster, and Curran, the insect specialist, referred to it as a "man from Mars" photo. In reality, he said, it was only a highly magnified image of a marsh springtail, a species of insect found all over Europe and Asia.

One of the alleged "poisonous insects," said Curran, was a mosquito whose wings had been removed. The other supposed killers were in fact stoneflies, "perfectly harmless creatures often found in large numbers within varying distances from water, in which the larvae live. Their flight is slow, and because they are poor fliers they would be extremely poor at disseminating disease even if they could be inoculated."

As for the paralyzed flies, "the fact that they crawled about but didn't fly means that they were too cold to fly. This happens to all flies when they become chilled." (The pictures had been taken outdoors, in winter.)

The shots of bacteria fared no better. The alleged "meningitis double globular bacterium," said René Dubos, the bacteriologist, was "not a meningitis bacillus. It is the wrong shape—round instead of kidney-shaped—and is single, whereas meningitis bacilli occur in pairs."

The "consecutive-globular bacteria," he said, were actually examples of *Microccus tetragenus*. "Everyone has them in his throat. They are not known to carry diseases."

The "bluish impurities," likewise, were "absolutely nothing. Some junk colored by dye. Meaningless as evidence, monumentally insignificant."

On the same day that the *Times* published its story of the Chinese picture hoax, the paper ran a separate news item about an epidemic of schistosomiasis that was then taking place in China. Schistosomiasis was an intestinal disease caused by parasitic flatworms. Next to malaria, it was the world's most common infectious disease, long known to have been widespread in the Far East, and it was currently infecting some ten million people in four East China provinces. Efforts to combat the disease, the *Times* story said, were hampered by a lack of health personnel.

"The greatest difficulty in our work is a shortage of cadres," said Kung Nai-chuan, a Chinese health official. "There are only ten doctors or so who are regularly in charge of prevention work and treatment." In some localities, he said, up to ninety-seven percent of the residents had contracted the disease.

The *Times* account of the outbreak lent some credibility to the claim, made by American officials at the time and long afterward, that the Chinese and Korean germ warfare charges were attempts to cover up their inability to deal with their own recurrent seasonal epidemics, while at the same time scoring needed propaganda points against the imperialist invaders from the West.

Hard on the heels of the democratic lawyers, however, a second group of "progressive" investigators was headed for the Far East. This was the International Scientific Commission for the Investigation of the Facts concerning Bacterial Warfare in Korea and China, otherwise known as the International Scientific Commission, or ISC.

The ISC had been put together by the World Peace Council, a Soviet-backed political group with headquarters in Oslo. Of the eight commission members, three were citizens of Communist bloc countries or openly sympathetic to the Communist cause: one was from the Soviet Union, another was from Communist China, all of them under the leadership of Joseph Needham, a Cambridge University biochemist and an avowed Marxist.

On June 23, 1952, the group rolled into Peking.

Over the course of the next two months, the members combed the trail blazed earlier by the democratic lawyers and searched for new evidence of germ warfare attacks. They investigated the case of the insects, the episode of the voles, and the tale of the clams, along with forty-seven other instances of alleged biological warfare in the Far East. They viewed documents, interviewed eyewitnesses, and studied photographs. They also met with four captured American fighter pilots while they were still in captivity, and listened to their remorseful germ warfare "confessions."

At the end of it the investigators came back and produced a 60-page text capped by 600 pages' worth of appendices, the formal and final conclusion of the whole being, "The peoples of Korea and China have indeed been the objective of bacteriological weapons. These have been employed by units of the U.S.A. armed forces, using a great variety of different methods for the purpose, some of which seem to be developments of those applied by the Japanese army during the second world war."

Despite the *New York Times* exposure of the Chinese picture hoax just three months earlier, and despite denials from United States military and civilian officials, it would be a long time before the Americans escaped from under the cloud of suspicion created by the ISC report.

Camp Detrick's *Serratia* assaults on the Pentagon, its large-scale mock attacks on the east coast of the United States and on the city of San Francisco, and the mass airborne disseminations of fluorescent particles over the rest of the country were all vulnerability tests and defensive in nature. Their object had been to determine the extent to which the country was open to covert biological attack. But such experiments constituted only a small portion of the SO (Special Operations) Division's assignment at Detrick. The major share of their work was offensive in nature, creating means and methods of waging covert attacks upon individuals or small groups.

There were two key requirements for such attacks: a biological agent that was appropriate to the job, and a delivery mechanism suitable for getting the agent to the target. Accordingly, the SO Division had two main sub-branches, the Agent Branch and the Device Branch. Chief of the Agent Branch was Ben Wilson; chief of the Device Branch was Herb Tanner, inventor of the 8-Ball, the largest biological device the world had yet seen.

It was the responsibility of the Agent Branch to select out from the world's infectious microorganisms a discrete subset of hot agents that was best suited to a variety of possible objectives: marking a person (with a rash, for example) without inducing sickness; inducing a mild or a serious illness; a long illness or a short one; instantaneous death, death after a time delay, death after a protracted sickness, et cetera. There were different hot agents for all these various functions, and it was the job of the Agent Branch to identify, test, and stockpile them for possible use.

The Device Branch, by contrast, came up with suitable delivery systems for a variety of different agents and circumstances. From the start, the Device Branch was the realm of classic secret-agent hardware, the stuff of thrillers and spy novels: poisoned chewing gum, coated missiles, fountain-pen dart launchers, attaché cases that spread bacterial aerosols, cigarette lighters that sprayed germs at the user, bacterial powders for dusting the victim's clothes, shoes, and pillows, engine head bolts that released deadly gases when heated, a "nondiscernible microbioinoculator" that penetrated the skin without the subject's knowing it, so on and so forth. All these gadgets and many more were developed and tested by the SO Division's Device Branch.

The CIA had obvious and pressing needs for such trick biological agents and mechanisms, not to mention a dependable suicide pill for use *in extremis*, and who better to provide these things than the experts? So in May 1952, the CIA contracted with the U.S. Army Chemical Corps whereby in return for an annual payment of $200,000, the Detrick SO Division would supply the CIA's Technical Support Staff (TSS) with various lethal and incapacitating germs and toxins, together with suitable dissemination systems, for possible use in clandestine activities against human targets. The TSS was the CIA's own in-house gadget factory, specializing in counterfeit documents, hidden cameras, radio transmitters concealed in false teeth, and the like. The CIA had official code cryptonyms for every project, and those within the TSS branch began with the two-letter code MK. The joint CIA-Detrick program was initially called MKDETRIC, a name whose meaning, the agents soon realized, was far too obvious. They later changed it to MKNAOMI, a name that apparently meant nothing at all.

As a later CIA document described it, Project MKNAOMI "was characterized by a compartmentation that was extreme even by CIA standards. Only two or three Agency officers at any given time were cleared for access to Fort Detrick activities. Because of the sensitivity of the activity, queries by operations officers as to the availability of materials and delivery systems of the type being developed at Fort Detrick were automatically turned away unless initial approval for contact had been given by the Deputy Director for Plans."

The Agency officer responsible for overseeing the Detrick-CIA connection was Sidney Gottlieb, a Caltech biochemistry Ph.D. who lived on a fifteen-acre goat farm and Christmas tree plantation in Vienna, Virginia. Gottlieb was the current head of the TSS. However, so very top-secret was the link between the TSS and the SO Division that when Gottlieb had a substantial amount of business to discuss with the Detrick scientists, they would go off for a secret meeting in the mountains a hundred miles or more to the west. Two favorite locations for these "retreats," as they called them, were Lost River, West Virginia, and Deep Creek Lake in western Maryland.

A year into the project, in April 1953, CIA director Allen W. Dulles created a second secret undertaking within the confines of MKNAOMI. The new program was MKULTRA, whose purpose was to discover, test, produce, and stockpile "chemical and biological materials capable of producing human behavioral and physiological changes." MKULTRA, in other words, was the CIA code name for mind control. In November 1953, Sid Gottlieb decided to perform an impromptu mind-control experiment at a retreat to be held over a three-day period at Deep Creek Lake.

Deep Creek Lake was three hours by car from Camp Detrick. On Wednesday morning, November 18, 1953, about a week before Thanksgiving, a group from the SO Division, including Vincent Ruwet, chief of the division, John Schwab, Frank Olson, Ben Wilson, Gerald Yonetz, and John Malinowski, drove out to the retreat.

The route went west through pretty mountain scenery to Keysers Ridge, then south through the town of Accident before fetching up at a rustic two-story log and stone lodge on the northern shore of the lake. It was late fall, almost winter, and the place was deserted.

The Detrick group was met at the lodge by Sid Gottlieb, his deputy Robert Lashbrook, and a couple of others from the CIA.

On the second day of the retreat, after dinner, Gottlieb spiked a bottle of Cointreau with a small quantity of a substance that he and his TSS colleagues privately referred to as "serunim" but which was in fact lysergic acid diethylamide, or LSD. It had been discovered in 1938 by two chemists working at the Sandoz pharmaceutical company laboratories in Basel, Switzerland, Arthur Stoll and Albert Hoffman, the latter of whom, five years later, accidentally ingested a bit of it while at work in the lab. He felt as though he were drunk. He was wobbly and dizzy, couldn't concentrate, and his mind seemed to race along of its own accord, producing a variety of fantastic colors, shapes, and images inside his head. Hoffman wrote up an account of the drug and its effects and published it in 1947.

Two years later, in 1949, a Viennese physician by the name of Otto Kauders gave a lecture at the Boston Psychopathic Hospital, a mental institution, about the drug's ability to make sane people temporarily "crazy." His idea was that, conversely, maybe the substance could also help to make the crazy sane. At all events, taking an experimental dose of it could afford mental health researchers a fleeting vision of what it was like to be mentally ill.

Word of the drug soon reached the CIA, whose agents regarded it, because of the way in which it wrested control of the mind away from the subject, as a valuable new espionage tool. Conceivably, an involuntary dose of it could unlock the minds of foreign spies and force them to cough up their most precious secrets. But the drug had to be tested first, and so the members of the Technical Support Staff, including Sid Gottlieb himself, started taking it on an experimental basis. Still, the fact that they *knew* they were taking it interfered with the results, and to get a clear picture of what the drug could do they wanted to observe its effects when taken unwittingly. To Sid Gottlieb the Deep Creek Lake retreat seemed like the perfect setting in which to perform just such an experiment. So he poured out the "serunim"-spiked drinks and offered them around to those present, most of whom accepted a glass.

Nothing happened. At least not for a while. But twenty or so minutes later, when Gottlieb asked if anyone noticed anything—*How are you feeling? Everything okay?*—many of them realized that some subtle changes had indeed taken place in their heads. Others, Gerald Yonetz, for example, seemed to have no reaction.

Gottlieb then told them what he'd done, a confession that clearly did not please the listeners. Frank Olson, in particular, thought that he'd been used, that Gottlieb had played a mean trick on them all.

There was no more talk that night of biological agents or devices. The meeting degenerated into laughter and silliness, and one by one people drifted off to bed.

Not that they could sleep. The morning after the experiment, most of the test subjects were still too muddleheaded to talk business. The gathering broke up and everybody left.

Dugway's M33/*Brucella* bombing run of August 9, 1952, had been followed by a second mission on August 28 over a fresh supply of 3,230 guinea pigs, a third on September 18 over 2,584 guinea pigs, and a final bomb drop on October 7 on 2,584 additional animals. By the end of the M33 operational suitability trials, a total of 11,628 guinea pigs had been attacked by *Brucella* bombs at Dugway Proving Ground within the space of two months.

As an Army Chemical Corps General remarked years later, "Now we know what to do if we ever go to war against guinea pigs."

The official verdict of the Air Proving Ground Command was that the *Brucella* clusters did not hold a candle to the atom bomb: "The M33 bomb cluster with *Brucella suis* compared with the atomic bomb would produce only ⅓ to ½ as many casualties in an attack on a typical target city, but would require 7 times as many bombers. The average loss of labor force during the first year after an attack would be 44 per cent for the atomic bomb, but only 7 per cent for a biological munition." (The Air Force, at one point, had briefly considered an atomic and biological bomb combination: "BW might produce its worst effects when employed simultaneously with the atomic bomb," said an Air Force secret document. "Theoretically, by proper time fusing of the 4-pound biological bomb, a ring of BW contamination could be established completely around the areas affected by the blast, heat, and radiation of an atomic explosion. Needless to say, the burden on medical services and civil defense organizations in that event would be intolerable. Demoralization would be complete if medical and public health organizations broke down under the impact of a combined atomic and biological warfare attack." Nothing ever came of the idea.)

Air Force strategists calculated, anyway, that in order to infect thirty target areas of thirty square miles each, some 17,000 M33 clusters, containing a total of 1,836,000 M114 biological bombs carried aboard 1,221 C-54 aircraft, or their equivalent, would be required. That was a lot of bombs, but not beyond the realm of possibility.

Still, weak though it was, the M33/*Brucella* munition deserved its own niche in a "balanced arsenal." While the A-bomb was far more efficient than biological bombs at producing lots of casualties in one fell swoop, there was no guarantee that the Air Force would be able to draw on atomic weapons in the future. The atom bomb might be the most efficient munition in the inventory but nevertheless be unusable in a particular circumstance for political, moral, or policy reasons. Biological weapons were really no substitute, but they represented a fallback position.

Also, they had an undeniable psychological impact. If they were less effective than atomic bombs, or even conventional high explosives, they could anyway be relied upon to scare a population out of its wits, perhaps terrorizing them into submission. For all these reasons, the Air Force was ready to make them a part of its arsenal.

On July 1, 1953, accordingly, the Air Force formally adopted Air Materiel Command Operational Plan 13-53. The plan outlined what would happen if and when *Brucella suis* clusters were ever to be sent into an overseas theater. Acting on direct orders from the president of the United States, who would authorize the devices only in retaliation for a biological attack, the Air Force would notify the Army Chemical Corps to begin crash production of *Brucella* suspension at their new bacterial production facility at Pine Bluff Arsenal, a follow-up to the abandoned Vigo plant. Workers at Pine

Bluff would fill a succession of M114 bombs with the *Brucella* and then gather the filled bombs into clusters. They'd load twelve refrigerated trailers with thirty M33 clusters apiece, and then they'd transport them by highway, followed by a specially trained decontamination team, to the designated aerial port.

At the aerial port, the trailers with the filled clusters inside would be put aboard a series of C-124 Globemaster transport aircraft that would fly them to one or more points in Europe. There the trailers would be unloaded from the aircraft, the clusters would be extracted from their refrigerated carriers, and then they'd finally be loaded aboard the bombers that would actually fly the strike missions. The bombers would fly off to the target area while the empty trailers were flown back to the "Zone of the Interior," a point in the United States from which the whole cycle would begin again. And in this fashion, without stockpiling any appreciable amounts of hot agent, the Air Force could mount more or less continuous biological bombing runs on demand, at the rate of 2,000 *Brucella*-filled M33 clusters a month.

In June 1953, the Air Force placed an order for sixty-seven refrigerated trailers ("semi-trailer, temperature controlled van, demountable") from Brown Trailers, Inc. That was the secret of biological munitions, apparently: refrigerators. The Air Force also ordered two additional laboratory trailers ("laboratory, field surveillance, portable, type MA-1"), and by August of 1953 the rigs started rolling into Wright-Patterson Air Force Base in Ohio.

The Army's new biological production plant at Pine Bluff Arsenal, meanwhile, was completed in December 1953 and was ready for operation by June of the following year. It immediately started producing basic startup amounts of *Brucella suis*, and soon was keeping enough of the agent on hand to fill 360 M33 clusters within twenty-four hours of the required seventy-two-hour advance notice. The standby *Brucella* was stored in holding tanks for a period of two to three weeks, after which, if none of it was needed for use in battle, it was sterilized by heating and then disposed of. Before disposal, however, fresh batches of the agent were grown, tested, and placed on reserve.

Although the *Brucella* itself would not be stockpiled except in these small, rotating lots, the necessary inanimate bomb casings, cluster adapters, and other pieces of relevant hardware would be produced in vast numbers, and within a year of Operational Plan 13-53 having taken effect, the Army had 2,581,200 empty M114 bomb casings in storage (enough for 23,900 M33 clusters), plus the production capacity to turn out an additional 2,376,000 bomblets (enough for 22,000 clusters) every month.

By mid-1954, then, with a hot agent mass-production line primed and ready to go, with millions of bomblets and tens of thousands of cluster adapters in storage, and with ranks of refrigerated trailers awaiting the arrival of filled munitions, the United States, finally and at long last, was equipped to wage biological battle.

When Vincent Ruwet, chief of the SO Division, arrived for work on the Monday morning following the Deep Creek Lake retreat and the impromptu LSD experiment, waiting for him in the vestibule of his office in Building 439 was Frank Olson, chief of the division's Plans and Assessment Branch.

Olson was thin and sandy haired, and he was a known live wire at Detrick. Ruwet was an aristocratic-looking Army scientist who bore a slight resemblance to the Hollywood actor Stewart Granger. The two had known each other since Ruwet had come to the SO Division in 1951. Both of them had gotten graduate degrees in microbiology from the University of Wisconsin, although they did not cross paths there.

Ruwet was surprised to see Olson at 7:30 in the morning, but asked him in. Olson told Ruwet that he was dissatisfied with his own performance at the retreat, that he was experiencing considerable self-doubts, and that in fact he had decided he would like to be out of the germ warfare business. He wanted to leave Camp Detrick and devote his life to something else. (Privately he toyed with the notion of becoming a dentist.)

Ruwet told Olson that as far as he was concerned Olson's recent work was just as good as it always had been, which was outstanding, and that his participation at the retreat had been entirely beyond reproach. Olson listened to this and went away reassured.

He was back in Ruwet's office again at 7:30 A.M. the next morning, complaining of confusion, inability to concentrate, sleeplessness, along with other signs and symptoms. He was, he said, "all mixed up."

Ruwet now decided that this was far more serious than he'd realized at first. Outside intervention was clearly advisable, not only for Frank Olson's sake, but for the sake of Camp Detrick and the biological warfare program, and in particular for the overall security of the SO Division. Everyone's worst nightmare had always been of someone's flipping out and running amok, and spilling all the family secrets.

Ruwet called Sid Gottlieb at CIA headquarters and told him of the problem. Gottlieb said there was a physician in New York City, a Dr. Harold Abramson, who was a cleared consultant of both the CIA and the Chemical Corps and who treated Agency personnel with psychological problems, particularly those stemming from experiences with LSD. Gottlieb suggested that Abramson see Olson, and so Ruwet, Olson, and Robert Lashbrook, Gottlieb's second-in-command, flew up to La Guardia Airport in New York that same afternoon, Tuesday, November 24, 1953. Olson was highly agitated and anxious during the trip; he again stated that he was "all mixed up" and said that he felt as if someone was out to get him, although he didn't know why.

On arriving they took a taxi from the airport to Abramson's office at 133 East 58th Street in Manhattan, an elegant redbrick townhouse with wrought iron grillwork at the windows. Harold Abramson had gotten his M.D. degree from Columbia University in 1923. During World War II he'd worked for the Chemical Warfare Service doing research on the use of aerosol penicillin therapy for various diseases, and he'd come into brief contact with Frank Olson at that time. Beginning in the early 1950s, Abramson did LSD-tolerance experiments for the CIA, and in 1959 he would edit a book called *The Use of LSD in Psychotherapy.* (Project MKNAOMI had in fact been named after Abramson's secretary, Naomi Busner.)

For a while Olson, Ruwet, and Lashbrook made social conversation with Abramson in his office. Olson remembered Abramson from their days at Camp Detrick together, and for a while they talked about protection devices, mutual friends, and the like. Then Ruwet and Lashbrook left and Abramson met with Olson alone.

Olson told Abramson that he was dissatisfied with himself and his work, that his memory was poor, that his spelling had deteriorated, and that he was in above his head at his job. So serious were his self-doubts, he said, that when he'd been appointed acting head of the SO Division he'd panicked and soon asked to be relieved of the position. He hadn't been able to sleep well since March, seven or eight months ago. He'd been experiencing guilt feelings on account of having been retired from the Army with a disability pension because of an ulcer. This was dishonest, he felt, even though he really did have an ulcer.

Ruwet and Lashbrook came back and picked up Olson at Abramson's office about 6 o'clock. The three of them checked into the Statler Hotel at

Seventh Avenue and 33rd Street, across from Pennsylvania Station on the west side of the city. The Statler was an enormous place, with some 1,700 rooms.

That night, Abramson visited Olson, Ruwet, and Lashbrook in their adjoining two rooms at the Statler. It turned into a festive occasion and Olson appeared calm and composed. As Abramson was leaving, Olson said: "You know, I feel a lot better. This is what I have been needing."

Next day, Wednesday, November 25, the three returned to Abramson's office, where Olson had a 4 o'clock appointment. Olson repeated his previous litany of complaints: poor memory, confusion, inadequacy, his work wasn't up to par. Abramson couldn't square Olson's claims of poor memory with the sheer volume of detail that Olson was recollecting right in front of him; he seemed to have an excellent recall of people, places, and events. Likewise Abramson couldn't harmonize Olson's own view of his job performance with the fact that he'd been appointed acting head of the entire SO Division. But none of this made any difference to Olson, who also reported that he'd been blurting out classified information as if he couldn't help himself.

Back at the hotel, Olson told Lashbrook and Ruwet that he'd failed in his job, that he was a disgrace to his colleagues, friends, and family, and that they should just forget about him and let him "disappear."

To get Olson's mind off his problems, Ruwet bought tickets to the Broadway hit musical, *Me and Juliet*, and the three of them went to the play. During the first act, Olson told Ruwet that people were outside waiting to arrest him at intermission. Ruwet and Olson left the show at that point while Lashbrook stayed on till the end.

They went to bed a little after midnight.

Then it was Thursday, November 26. Thanksgiving.

Ruwet awoke at 5:30 in the morning and saw that Olson's bed was empty. He woke Lashbrook and the two of them searched for Olson, whom they found sitting in the lobby dressed in his overcoat and hat.

He'd been awake since 4 A.M., he told them. He'd been wandering the streets. He'd thrown away his personal identification papers, ripped up his money, and tossed his wallet down a chute. He'd done all this, he said, under the direct orders of Vincent Ruwet.

Olson had wanted to be home with his family for Thanksgiving, so Ruwet and Lashbrook flew back with him to Washington. On the drive from Washington out to Frederick, Olson changed his mind about going home. He couldn't face his family, he said. Instead, he wanted to go back to New York and see Dr. Abramson.

Abramson, however, was at his home on the north shore of Long Island having Thanksgiving with his own family. Still, he would see Olson if he wished, and so Lashbrook and Olson flew back to New York that same afternoon and drove out to Long Island. Ruwet, meanwhile, stayed behind in Frederick.

Abramson lived at 47 New Street in Huntington, and Lashbrook and Olson arrived at the house about 4 P.M.

Olson now told Abramson that the reason he couldn't sleep was that the CIA was putting something in his coffee, maybe Benzedrine . . . People were trying to get rid of him . . . He was hearing voices . . .

Abramson recommended that Olson be hospitalized, and Olson did not reject the idea.

Lashbrook and Olson spent the night at the Anchorage Guest House in Cold Spring Harbor. It was exactly a week since he'd been given the LSD at Deep Creek Lake.

Friday morning, Abramson picked them up at the guest house and drove them back to his office in the city. After speaking with Sid Gottlieb by phone, Abramson arranged to have Olson admitted to Chestnut Lodge, a sanitarium in Rockville, Maryland. This was satisfactory to Olson.

On their final night in the city Olson and Lashbrook stayed once more at the Statler, where they were given room 1018A. The "A" referred to the northwest wing of the building. The room had two double beds, a television, and a window that overlooked Seventh Avenue. Because of the way the hotel's floors were numbered, room 1018A was actually located thirteen stories above ground level.

The two went into the cocktail lounge where they each had a couple of martinis. Olson spoke of his coming hospital stay and told Lashbrook how he planned to spend his time reading books and doing various scientific projects. To Lashbrook, Olson seemed happy enough at the prospect.

They had dinner in the hotel restaurant, the Café Rouge, then they went up to the room and watched television.

Olson spoke by telephone to Vincent Ruwet in Frederick, telling him he'd see him in a day or two. He then called his wife, Alice, for the first time since he'd come to the city. He told her not to worry, that he'd be seeing her soon. He washed out a few of his clothes in the bathroom sink and left a wake-up call with the hotel operator so that they'd be sure to make their flight the next morning.

Just in case Olson should try to leave the room and wander about the neighborhood as he'd done two nights earlier, Lashbrook took the bed next to the door.

Now, around midnight, they went to bed.

Only ten days previously, Frank Rudolph Olson, Ph.D., had been a branch chief in the Special Operations Division, a trusted and valued employee of the U.S. government's secret germ warfare installation at Camp Detrick. Now he was hearing voices, having delusions, and on his way to the crazy house.

He couldn't get to sleep, couldn't stay asleep.

At about 3 A.M., with Lashbrook asleep, Frank Olson crashed through the closed window shade and the closed window of room 1018A and disappeared below the ledge.

part three

12

It was a shortcoming of the American biological warfare program that even after the Air Force put Operational Plan 13-53 into effect and the Army started stockpiling bomb casings, cluster hardware, standby supplies of pathogen, and refrigerated trailers, neither the Air Force, the Army, nor anyone else in the Western world had any firm notion as to what a drop of *Brucella* bombs would do to the human population of an enemy city. With the exception of the bubonic plague inoculations of San Quentin prisoners during World War II, all the germ warfare experiments that Western researchers had ever done had been performed on animals used as test subjects. The problem, however, lay in extrapolating from animal data the likely effects of the same biological agents on humans. As a U.S. Air Force historical document expressed it, "The Air Force could be fairly accurate in predicting what a biological warfare attack would do to a city full of monkeys, but what an attack would do to a city full of human beings remained the 'sixty-four dollar question.'"

The Army tried to answer the question in two different ways, one theoretical and one empirical. In the theoretical approach, the Chemical Corps contracted with a group of scientists at the University of Pennsylvania and asked them to prepare a mathematical analysis of the animal-human extrapolation problem. This was part of a larger effort code-named "Project

Big Ben," a broad attempt to assess the overall workability of biological munitions. Under a narrow subprogram called "Project Ball Game," the University of Pennsylvania scientists developed a computational model, on the basis of which they hoped to be able to generalize from the effects of a given microorganism on a succession of different animal species to the probable consequences of that same microorganism upon humans. They plotted the results of thirty-eight kinds of pathogens on twenty-one separate animal species. From observing the effects on species that most closely resembled human beings, they estimated the probable result of those same pathogens on *Homo sapiens.*

This was an approach that held some promise. Still, due to the fact that they were unable to check their extrapolated predictions against experience, the scientists had no way of knowing how accurate their conclusions were. Theory, in other words, went only so far. Scientific knowledge of human susceptibility to biological agents would have to come from experience, from direct tests of those agents upon their intended subjects. In the fall of 1954, finally, Army medics seriously considered the idea of subjecting human beings to live biological warfare agents. All they needed was a supply of soldiers who would volunteer to be deliberately infected with pathogens.

The use of human volunteers was nothing new in medicine. Indeed, it was a standard and respected practice. New drugs, vaccines, diets, surgical procedures, machines, and other medical advances always had to be tried out by an initial wave of human guinea pigs before being released for use by the general public. Basically any drug that you could buy at the pharmacy had been previously tested by ranks of human beings who had given their full and informed consent to the experiment. Pathogenic biological agents were different from drugs in that they would induce sickness rather than cure it, but so long as the subjects were informed of the risks involved and consented to them beforehand, what was the problem?

In the U.S. Army there was a group of soldiers who, because of their religion, were noncombatants, but who, also because of their religion, took better than usual care of their bodies: the Seventh-day Adventists. These unusual specimens didn't smoke, didn't drink, and didn't use caffeine, and most of them didn't eat meat. In October 1954, Lieutenant Colonel William D. Tigertt, a physician at the Walter Reed Army Institute of Research, contacted Theodore R. Flaiz, M.D., of the Seventh-day Adventist Medical Department in Washington, D.C., and presented him with what he described as an unparalleled opportunity for American Seventh-day Adventist soldiers to assist in the national defense.

Although germ warfare had never been waged upon the United States, said Tigertt, it nevertheless remained an ever-present threat. To protect American soldiers against such an attack, the Army Medical Corps was developing vaccines and treatments for a wide range of biological agents. However, the Army was hampered by an inability to test those agents on humans, the intended targets of biological assault. This was where the Army's Seventh-day Adventists came in. If the soldiers would allow themselves to be infected by incapacitating microbes in a highly controlled experimental setting, then they would be contributing both to the national defense and to public health.

Theodore R. Flaiz welcomed the proposal, for the members of his church had a long and strong tradition of cooperating with the military. The Adventist religion dated back to the late 1830s, when a Baptist preacher and former Army captain by the name of William Miller predicted the second coming, or "advent," of Christ, an event that he foretold would occur on October 22, 1844. Unfortunately it did not, and this historic failure soon became known among his followers as "the Great Disappointment." No matter: Christ was still coming, said Miller, and would get here sooner or later. Meantime, the church faithful would have to watch and wait and keep themselves prepared for the grand arrival.

The Seventh-day Adventist faith emerged and began to grow, mainly in New England and upper New York State. Its disciples believed in the Ten Commandments, especially in the one that said "Thou shalt not kill," a dictum that made for some problems during the Civil War. Although the Adventist church was against killing, fighting, or bearing arms, it was in favor of obeying the law of the land, believing that civil government itself was an institution ordained by God. It was proper for Adventists to serve in the Army, but only as noncombatants. Since the Adventists saw the human body as the repository of the holy spirit, the most logical role for an Adventist soldier to play during armed conflict was in caring for the sick and wounded. The Adventists operated a few of their own private colleges, and some of the schools even went so far as to offer pre-induction medical training, the better to prepare their Christian soldiers to serve both God and country in wartime.

By the mid-1950s, then, the Adventist cooperation with the U.S. armed forces was well established, and so when the U.S. Army Medical Corps decided they needed some human guinea pigs to assist with the national defense, the Adventist hierarchy could hardly say no.

"The type of voluntary service which is being offered to our boys in this research problem offers an excellent opportunity for these young men to render a service which will be of value not only to military medicine but to public health generally," Flaiz wrote back in reply to Tigertt's request for volunteers. "It should be regarded as a privilege to be identified with this significant advanced step in clinical research."

Thus began Project CD-22, also known as "Operation Whitecoat." Over the next twenty years some 2,200 U.S. Army Seventh-day Adventists would volunteer themselves up to be infected with a variety of the Army's offensive pathogens, chief among which were the causative agents of Q fever, tularemia, sandfly fever, typhoid fever, Eastern, Western, and Venezuelan equine encephalitis, Rocky Mountain spotted fever, and Rift Valley fever.

The first canonical test, with the Q fever microbe streaming into the open air and washing over the test subjects, would take place at Dugway Proving Ground in the Utah desert, on July 12, 1955.

At Fort Sam Houston in Texas, a medical training center where many of the Seventh-day Adventists were stationed, Wendell Cole, Lloyd Long, Louis Canosa, Richard Miller, and some others watched as Colonel Tigertt himself made the pitch for the Q fever experiment. Tigertt was tall and thin and wore a pencil mustache, and seemed to be entirely open and aboveboard about the benefits and risks involved.

Q fever, he told them, had been discovered in Australia in the 1930s, but it had also affected a thousand American troops in Europe during World War II. It therefore constituted a significant military threat, and the Army wanted to understand the agent's infectiousness when used as a weapon and to develop antibiotics and vaccines against the disease. That's what the volunteers were needed for.

Anyone who was accepted into the program would come up to Maryland, the home of Camp Detrick, the Army's top secret germ warfare research facility. You might end up being exposed to the microbe or you might not be: you wouldn't know in advance, as most of the trials would be conducted on a double-blind basis. Also, you might be immunized prior to a given exposure, by inoculation or by other means, but this was not guaranteed, since nonimmunized subjects were needed as controls. Anyone exposed to the microbe without previous immunization would probably get the disease, but competent medical treatment would begin as soon as the first symptoms appeared. Q fever was normally not fatal, and with immediate treatment and the proper drugs the chances of anyone dying were extremely remote.

You could change your mind at any point in the proceedings, said Tigertt, with no penalties whatsoever. If you went ahead with the project, there would be no monetary payment involved, but in return for your participation you'd be able to choose the location of your future tour of duty. Anywhere in the world was fair game.

To Cole, Long, Canosa, and Miller, all this sounded mighty interesting. They went to the post library and looked up Q fever, learning not much more than what Colonel Tigertt had already told them. There was no mention of its possible use in biological warfare, but it was somewhat fascinating to contemplate being subjected to such an exotic disease in a controlled experiment. Plus they'd be helping their fellow soldiers, and they could write their own ticket in the future. All four of them volunteered for the Q fever trials.

Soon some thirty Seventh-day Adventist volunteers were flown up from Texas to Andrews Air Force Base, near Washington, D.C., and then brought by bus to their temporary residence at Forest Glen, a place that, in and of itself, constituted a serious departure from ordinary reality.

From the turn of the century until the late 1930s, Forest Glen had been the home of the National Park Seminary, a finishing school for young society ladies. It was the most expensive such place in the country, and every inch of the grounds reflected the fact. The main building contained parlors, an art studio, a social hall, fireplaces, and a dining room where 400 students at a clip were served by fifty uniformed waitresses. The gymnasium building nearby was in the Greek revival style and resembled the Parthenon; inside there were bowling alleys, a running track, a solarium, a pool, and "needle baths," which was the refined term for showers. Across from the gymnasium there was an athletic field with bleachers, tennis courts, and a set of swings. Not far away were the stables, for many of the young ladies brought their own horses with them.

There were, in addition, eight sorority houses, each built on a different architectural theme. There was a wood-shingled Dutch windmill complete with fan blades. There was an English stone castle with crenelated turrets and a working drawbridge that could be raised and lowered by means of chains. There was a Japanese pagoda, a Spanish mission, and a Swiss chalet. Off in its own separate space was the main dormitory building, an Italian-style villa surrounded by terraced gardens, fountains, and statuary.

The capstone of the place, however, was Ament Hall, designed by the school's headmaster, Dr. James E. Ament, an amateur dentist and self-taught architect whom the girls liked to refer to as "Dr. Dement." Ament Hall was a tall Gothic cathedral-like structure whose vaulted ceiling and stained glass clerestory windows gave it the outward appearance of a medieval church. The cavernous space inside it held no pews, however, for it was not a house of prayer but rather a ballroom in which the formally gowned ladies danced with each other after lunch and dinner. Bordering the ballroom on the upper floors were a number of two-room suites occupied by the wealthier girls and their maids.

During World War II, when the Medical Corps needed space for the wounded, the U.S. Army suddenly came to Forest Glen, formally condemned the National Park Seminary, and purchased the grounds, buildings, and furnishings for $855,000. They proclaimed the site an annex of the Walter Reed Army Medical Center and converted it into a recuperation facility for amputees. The main dining room became a mess hall, the gymnasium became a movie theater, and, following the inflexible rule that wherever there was Army there was Ping-Pong, the ballroom became the Ping-Pong room.

Despite these comforts and attractions, the recuperating war victims hated the place, the reason being that the architecture only reminded them of the bombs and gunfire of Europe and East Asia. The Japanese pagoda, the Italianate villa, the Swiss chalet, the British castle, and the Dutch windmill collectively made it seem as if they'd never left the war zone.

The amputees departed at the end of the war, at which time the place was given over to psychiatric cases, for whose benefit padded cells and barred windows were installed. After a while they, too, fled the scene, leaving Forest Glen largely forlorn and abandoned.

Into this surreal landscape there now arrived a bunch of young and innocent Seventh-day Adventists from Texas who took up residence in the Main Building and Ament Hall and made themselves at home. Or tried to.

Every so often the whole group or selected members would be bussed from Forest Glen to Camp Detrick where they'd undergo some medical procedure or other. Having blood drawn was the least of it. By the end of the program the volunteers were drawing each other's blood once a day, sometimes twice a day, and all of them got to be experts at it. Perhaps you'd get an injection, another time you'd be exposed to the Q fever agent, or maybe a simulant, or even just air—you never knew what—at the 8-Ball.

The first time Louis Canosa entered Building 527, which housed the 8-Ball, he looked up at this swollen silver object overhead and he said to himself: *My God, this is a big baseball! A giant grapefruit!*

They took him up in the elevator to the equator level, gave him a white lab coat—you always wore a white lab coat for any procedure, from which custom the men began to be referred to as Whitecoats—and led him into a telephone booth–type enclosure where instead of a phone there was a rubber hose that terminated in a face mask. The rubber hose went straight through the side of the 8-Ball and at the prescribed moment you put the face mask on, a technician flipped a switch in the control room and suddenly you were breathing the inner contents of the 8-Ball, whatever they were. It lasted only a minute and then you were done.

At that point, however, you were a hot object, you could be exhaling Q fever microbes out at all and sundry, and so Louis Canosa was immediately put into quarantine. He spent the next few weeks in a separate wooden hospital barracks at Detrick, one that had been reserved especially for the Whitecoats. There was television to keep him amused, plus reading materials, paint-by-numbers, and of course Ping-Pong.

Louis Canosa, however, never got sick.

After weeks of this, and what seemed like millions of medical tests and procedures, everyone was immunized who was going to be immunized, which was ten out of the thirty volunteers. It was finally time to go out to Dugway for the real test, the one true biological warfare trial out there in the American desert.

The men left from Forest Glen on the morning of July 5, 1955. The previous day, July 4, George DeMuth, one of the three Army physicians attached to the Whitecoat project, had given the volunteers yet one more physical check, turning up no big problems: a headache, a runny nose, the usual. Then that night everyone had gone into the city for the fireworks display over the Mall and the Washington Monument. Lloyd Long had his girlfriend with him as did many of the others, for while as a Seventh-day Adventist you were forced to abstain from cigarettes and alcohol, female company was nevertheless allowable. Besides, these were by definition healthy young males.

Tuesday, July 5, was a hot, clear summer day, and when the men got to Andrews Air Force Base, George DeMuth lined them all up in the sun in front of the plane for a group portrait. There were Wendell Cole, Ervin Bradburn, Elijah Foster, William Twombly, Richard Miller, Lloyd Long, Louis Canosa, George Parrish, Horace Beaty, plus some twenty others, all of them starched and spiffy in their dress khakis. Then they went up the stairs and onto the plane, a two-engine Military Air Transport Service Convair.

It was bumpy en route and Ervin Bradburn went up to the cockpit for a while and spent some time talking with the pilots. They stopped at Offutt Air Force Base in Omaha to refuel, and then a few hours later they landed at Dugway Proving Ground.

Dugway was out in flat, open desert, a place with no trees anywhere in sight, just some low juniper bushes, sagebrush, greasewood, and desert grass. The Dugway Mountains, named for the passageways the early pioneers dug in the slopes to get their wagons across, was a low range off to the south.

The actual test was to be held the following night. Biological trials took place after sundown, they knew, because the air was calmer then and stayed close to the ground as opposed to all the updrafts and turbulence of daytime. Also, sunlight instantly killed many pathogens, making a daylight test next to worthless.

So just before sunset the thirty Whitecoats piled into the back of a canvas-covered troop truck and watched as the shadowy landscape receded from view. Every so often there'd be a security checkpoint and an MP would come into the truck and look under the seats for cameras or other bootleg. George DeMuth, who'd taken the group picture at Andrews, now had his camera impounded by a guard.

The truck left the road and traveled for a while across the alkali flats, setting up a dust storm in its wake. When they finally got to the test area the desert floor was a dull blue and the smaller physical features had faded away.

They got out of the truck and looked around. They could see that they were completely surrounded by mountains except toward the northwest, which was an open stretch of flat desert punctuated by sand dunes. There was nothing else in any direction, no lights, roads, buildings, telephone poles, nothing. It seemed like the loneliest spot in the world.

In the near distance, however, they could see the test grid, a line of wooden platforms and tall stools of the type you might find at a kitchen counter or lab bench. The platforms and stools stretched off in a straight line that went on for more than a half mile across the crusted sand.

Nearby there was a rack of open showers fed by a water tank on the back of a flatbed truck. The men now stripped, showered, and got into clean clothes. Each of them was handed a blanket and then they walked to their places on the test grid, the ground around them crunching underfoot.

Lloyd Long, a nineteen-year-old recruit from Spokane, Washington, took up his assigned post and saw that on the platform next to him was a cage of seven or eight rhesus monkeys, all of them confined in metal boxes so that only their heads were exposed to the air. There was a second cage of boxed guinea pigs on the stand, plus an array of glass sampling devices and a vacuum pump to draw the air into the samplers.

Also at his post were two other men. Each person had his own separate chair, and all of them had been told to sit down and face the spray generators when the test began. The disseminators were a half mile away, over toward Granite Peak, so all they had to do when the time came was face the mountain. Supposedly, the generators would start spraying out the microbes soon after the drainage wind came down off Granite Peak.

The drainage wind, indeed, was what tonight's operation hinged upon, as it would control the direction and final fate of the pathogen cloud. During the day the sun heated the desert floor, causing currents of air to rise, whereas at night the reverse happened: the ground cooled, and the air drained down the mountain slopes and then blew across the salt flats where, eventually, it stalled and died out. The plan was for the drainage wind to carry the cloud to the empty salt flats and then stall out there and mill around until morning. Next morning, the sun would rise and kill any of the Q fever microbes that had managed to remain alive during the night.

Still, to be on the safe side and to determine whether any live agent escaped from the proving ground, the scientists set up a distant line of test animals along U.S. Highway 40. Route 40 was a full thirty-five miles from the dissemination point, a considerable distance, but there at the so-called peripheral sampling stations along the highway were an extra hundred or so guinea pigs, boxed and in cages like the rest.

Now the ground was cooling down and there was a perceptible chill on the night air. Lloyd Long wondered if the monkeys on the platform beside him were getting cold. He thought about draping his blanket over the cage but he wrapped it around himself instead. Soon it was totally dark.

And finally the wind started coming down from Granite Peak, a light cool breeze. It seemed as if the test would begin anytime.

But then the wind dropped, and then it stopped blowing entirely. Nothing more happened for an hour or so. And then the people in the control van canceled the test for the night.

Everyone abandoned their posts, went back to the decon station, changed into their old clothes, and returned to the barracks.

The next night was a repeat of the first. They showered in, got into coveralls, grabbed a blanket, and took up their positions on the test grid. Near midnight, they canceled the test.

The third day the group was bussed into Salt Lake City for a field trip. They went to the Mormon Tabernacle, they walked around Temple Square, they drove through the exceptionally wide streets and avenues of Salt Lake City, and then they left for the ninety-mile trip back to Dugway.

And again, the drainage winds did not exist.

On July 12, after almost a week of false starts, conditions were perfect: a gentle breeze, light and steady, from the direction of Granite Peak.

Colonel Tigertt, the man who'd gotten them all into this, walked down the rows saying, "Remember, when you hear the vacuum pumps, breathe normally. *Just breathe normally.*" Tigertt walked back toward the control van and disappeared.

The men heard the motors a few minutes later, the sound of the vacuum pumps drawing air into the samplers. They breathed normally. And almost before they knew it the test was over.

This time when they got back to the decon station and showered out, however, the procedures were a bit different from before. The test crew members in their hot suits plus goggles and face masks collected their coveralls and blankets, then brought them to an incinerator and burned them. The Whitecoats got new fatigues, and then they went from the test grid directly to the airport. On this trip, for the first time, even the driver was dressed like a spaceman. The security checkpoints were closed on the drive back, and no MPs were anywhere in sight.

As the airport came into view, Wendell Cole wondered how he'd ever get his old clothes and things back from the barracks. All his belongings, however, were already on the plane.

On the flight back, the pilots never once emerged from the cockpit, nor did they open the door to the cabin. At Andrews, the plane taxied to the far end of the airport, away from everything else, where a bus was waiting. The driver wore a white surgical mask. Soon they were at Camp Detrick.

Q fever hit Richard Miller as he was pushing an electric floor polisher around the dayroom of the quarantine building. The place was nice enough and there was plenty to keep you busy: Ping-Pong, crafts, and such. It was like being in summer camp. Lloyd Long would finish two paint-by-numbers horses' heads during his four weeks in isolation. But you could also do normal maintenance work, which was why Richard Miller was working with the floor buffer when all at once and without much warning he lost his strength, keeled over, and collapsed onto his own highly polished floor. That was the last he remembered until he woke up in bed.

At Dugway Miller had been near the middle of the test grid, in that part of the line known as the "dense sampling array," where the men, monkeys, and guinea pigs were all bunched up together. The pathogen cloud had apparently passed very nearly through the center of the dense sampling array because many of the others in that portion of the line were sick, too. Elijah Foster, who'd been seated within ten feet of Miller, was now in bed in his room across the hall—everyone had their own private room while in isolation—and he was not doing too well, which was strange, because Elijah Foster had been healthy, strong, and muscular. Lloyd Long, by contrast, who'd been positioned out toward the fringe of the test grid, spent only a single day in bed and then was up and walking around again.

Others, most probably those who'd been immunized beforehand, weren't sick at all—Wendell Cole, for example, never felt a thing. He'd had a first exposure at the 8-Ball, and that, evidently, had made him impervious to the disease. Louis Canosa, amazingly, never got sick either at the 8-Ball *or* at Dugway.

Still, they were all in isolation, and Dr. DeMuth came around every day and examined them one by one. It was he who officially decided whether you were sick or not, the criterion being whether you had a persistent fever above 100°F. If you did, then he'd give you a course of antibiotics, starting with a big loading dose of oxytetracycline, and you'd feel better within a day or two.

Miller had been in bed for three or four days with a fever and headache plus a weird choking sensation, and he didn't seem to be getting better. But then one day Malcolm Crofoot, who'd never gotten sick, left his record player going in his room down the hall and then went off to play Ping-Pong or something. The machine was set to play the same record, "Mood Indigo," over and over, automatically, and after approximately the twenty-third consecutive repetition, Richard Miller got up out of bed and went into Crofoot's room and smashed the thing to bits. That was when he knew he was recovering.

Nobody died from the Dugway Q fever trial. Nobody got permanently sick or incapacitated. Nobody suffered a relapse of the disease. The experimenters learned that larger doses of the Q fever agent shortened the incubation period, that previous vaccination prevented the disease, and that the infection was highly responsive to oxytetracycline.

Most important of all, the biological warriors now had their first eyewitness proof that aerosolized biological agents could infect not only caged lab animals but also live and healthy human beings, and that they could do so stealthily, silently, and from a distance of 3,200 feet.

That news was hardly surprising: it was exactly what all their animal tests had led them to expect. Nevertheless, they now had that expectation confirmed scientifically.

They'd also contained the Q fever microbe on the salt flats precisely as they had hoped to. The drainage winds had died out on schedule and the microbe had never reached a single one of the guinea pigs at the peripheral sampling stations along Highway 40.

As a precaution, the police at both ends of the highway had stopped cars and trucks and told the drivers to keep their windows shut and not to stop or get out for the next twenty-five miles or so. The test crew personnel out there at the periphery had had some fun with this while waiting for sunrise. One of the rubber-suited, masked, and goggled Dugway staff members had rolled around in the mud at the side of the highway. Then, as a car came toward him he loomed up out of the ditch and into the headlight beams, and started waving his arms like a crazy person. The cars speeded up and got out of there fast.

By the start of 1956, Camp Detrick had been in existence for thirteen years. It was acting like a fixed branch of the military rather than a temporary outpost in the hills of Maryland, and in formal government terms what this meant was that the name "Camp Detrick" was no longer appropriate to its station in life. A "camp" was a fleeting presence that could be abandoned overnight, a description that no longer applied to the installations at Detrick. On February 3, 1956, accordingly, "Camp" Detrick officially became "Fort" Detrick.

Six weeks later, the United States also made a slight change to its policy regarding the use of biological weapons. Driving the germ warfare project from the start had been the idea that, as the WBC Committee members had expressed it in their final report of June 1942, "the best defense is offense and the threat of offense." Coupled with this had been the notion that the United States would use biological weapons only in retaliation against a biological attack by the enemy. Since the country had no operational germ weapons with which to retaliate, its no-first-use policy had little practical bearing on the actual conduct of warfare. Still, the nation's military planners began to feel unduly restricted by the policy, which they thought inhibited preparedness. They'd be far better off if they were free to utilize any given weapon if, as, and when they saw fit.

In the early 1950s, as the prospect of an operational biological weapon drew nearer, the military planners sought a policy change, and by March 15, 1956, they were successful. On that date the National Security Council adopted regulation NSC 5062/1, a directive that stated, in part: "To the extent that the military effectiveness of the armed forces will be enhanced by their use, the United States will be prepared to use chemical and bacteriological weapons in general war. The decision as to their use will be made by the President."

Biological weapons, in other words, could now be used whenever the president authorized it, whether in retaliation for a biological attack or as a first strike.

13

frank Olson's head banged the top of the window frame as he crashed through the closed shade and window of the hotel room. It was a glancing blow, not enough to stop him, and he came out into the open air 175 feet above the sidewalk of Seventh Avenue, falling head first, arms flung out ahead of him, legs trailing, like a parachute jumper in free fall.

There were no abutments or projections on the building's facade but there was an eight-foot-high temporary construction fence on the ground near the hotel entrance and Olson's right arm smashed into it, the collision breaking his arm and two ribs. His body swiveled downward so that he hit the ground in a standing position and then fell backward, the right side of his head slamming into the sidewalk. The impact fractured Olson's right leg, his pelvis, and the right side of his skull.

He was flat on his back but still alive when the hotel's night manager, Armand Pastore, got to him. His eyes were open and he made a few sounds, but Pastore could not make out any words.

Olson stopped breathing and died a few moments later. Pastore now turned and looked overhead. About halfway up the side of the hotel he could see a window shade sticking out of a window. He counted up from the bottom and over from the left side of the building and figured out which room it was. By this time the police had arrived, and he and they took the elevator to the tenth floor and entered room 1018A with a passkey.

The lights were on inside. In the bathroom to the right a man in his underclothes was sitting on the toilet seat with his head in his hands. This was Robert Lashbrook, the CIA agent.

Armand Pastore went to the window. The glass was gone except for some fragments, and the shade was hanging out through the frame. He stuck his head out and looked down. Far below he saw the body of Frank Olson in his white undershirt and shorts, surrounded by a crowd.

Detective James W. Ward of the Fourteenth Police Precinct soon arrived in the room and started questioning Lashbrook. He'd been awakened by the sound of breaking glass, Lashbrook told him. He'd switched on the light next to him, seen the empty bed, and then saw the broken window and the shade hanging out of it. At that point he called the hotel operator. That was his whole story.

A couple of uniformed patrolmen searched the room but found nothing unusual. Lashbrook was giving no more than monosyllabic responses to further questions and Detective Ward wondered if Lashbrook and Olson had been engaged in a homosexual affair. The possibility of homicide could not be dismissed, however, and so Ward took Lashbrook in for questioning.

The station houses of the New York City police force were built like castles, their Gothic gray facades topped by balconies and turrets, and the Fourteenth Police Precinct at 138 West 30th Street was no exception. In an interrogation room, Ward asked Lashbrook to empty out his pockets. They contained airline tickets for Lashbrook and Olson between Washington and New York, a receipt for $115 dated 25 November 1953, and marked "Advance for travel to Chicago," a hotel bill, a postcard bearing the name and address of Vincent Ruwet, and papers bearing the name, address, and phone number of Dr. Harold Abramson and of the Chestnut Lodge in Rockville, Maryland. In the compartments of Lashbrook's wallet were various government passes and ID cards, and a small slip of paper printed with thirty unrelated letters of the alphabet; Lashbrook said they were a coded safe combination.

Lashbrook now identified himself as a chemist employed by the War Department. He told them that Frank Olson was a bacteriologist stationed at Camp Detrick, Maryland, that he was mentally ill, and that he'd brought Olson to New York to be examined by Dr. Abramson. The story ended with Olson's crashing through the hotel window.

He also told the detectives that all of this was extremely hush-hush and sensitive and that it would be best for everyone if they didn't look into it too closely. The detectives, nevertheless, checked with Abramson and Ruwet, both of whom gave them essentially the same version of events.

Later that morning, at the morgue of Bellevue Hospital at 29th Street and First Avenue, the assistant medical examiner of New York, Dominick DiMaio, performed an autopsy on the body of Frank Olson, finding that the death had been caused by multiple fractures, shock, and hemorrhage consequent upon a jump or fall from the tenth floor of the hotel.

On November 30, two days after the incident, Detective Ward wrote up a report summarizing the results of his investigation into case number 125124, the death of Frank R. Olson. All of Ward's initial doubts and suspicions had been banished as a result of his inquiries, and he ruled the death a suicide and closed the case.

The CIA now turned to the task of providing "employment backstopping" for Robert Lashbrook, which was to say, a suitable cover identity. Detective Ward had written in his report that "Due to the importance of the positions held by the deceased and LASHBROOK with the U.S. Government, the facts in this case were related to F.B.I. Agent GEORGE DALEM (by telephone)." The last thing the CIA wanted was for the FBI to be poking its nose into the Lashbrook-Olson business. In addition, Alice Olson had filed an insurance claim for double indemnity in connection with her husband's death, which meant that insurance investigators were sure to follow. For these reasons, it would be necessary to construct a fake employment history for Lashbrook, whose connection with the CIA had to be kept secret. The Agency maintained an entire department, the Cover Branch, to generate such misinformation, and so late in December 1953, one of the branch operatives met with Lashbrook to devise an appropriate cover tale.

In the interview, Lashbrook said that he wanted to be backstopped with Colonel Bjarne Furuholmen of Army G-4, essentially the Army's supply and logistics division. But the chief of the Cover Branch, Robert W. Cunningham, did not much like that idea. "Backstopping through the Logistics Office of the Army would be inconsistent with Dr. Lashbrook's background [Lashbrook had a Ph.D. in chemistry] and might create undue interest insofar as an insurance investigation is concerned," Cunningham wrote in a memo. Cunningham preferred that Lashbrook be backstopped at Camp Detrick, but Lashbrook was dead set against that since he thought it would only open a can of worms. To resolve the impasse, the Cover Branch agent paid a visit to Lashbrook's boss, Sidney Gottlieb, the man who had put all these events in motion to begin with by putting LSD into the drinks at Deep Creek Lake. But Gottlieb, too, was in favor of the G-4 cover story and wanted nothing whatsoever to do with any Camp Detrick connection.

And so it was done: "Cover arrangements for Subject were laid on with Col. Furuholmen and also with his deputy, Lt. Col. Jackson Lawrence," Cunningham wrote. "The cover story which was arranged for was that Subject is employed as an intermittent consultant by Col. Furuholmen's office and has been in this capacity for approximately 2-½ years. The nature of his duties are chemical research on classified material."

There now remained only the matter of Sid Gottlieb himself. If it hadn't been for him and his LSD experiment at the SO Division retreat, none of this ever would have happened. And so in February of 1954, Allen W. Dulles, the director of the Central Intelligence Agency, looked over the growing batch of documents pertaining to the Gottlieb-Lashbrook-Olson case and decided that something would have to be done. And so he sent Sid Gottlieb an official reprimand:

Feb 10 1954

PERSONAL

Dr. Sidney Gottlieb
Chief, Chemical Division
Technical Services Staff

Dear Dr. Gottlieb:

I have personally reviewed the files from your office concerning the use of a drug on an unwitting group of individuals. In recommending the unwitting application of the drug to your superior, you apparently did not give sufficient emphasis to the necessity for medical collaboration and for proper consideration of the rights of the individual to whom it was being administered. This is to inform you that it is my opinion that you exercised poor judgment in this case.

Sincerely,

Allen W. Dulles
Director

■   ■   ■

Poor judgment was not a bar to advancement in the CIA, however. By the start of 1960, Gottlieb, in addition to his other duties as head of the technical services staff, was also special assistant to the CIA's Deputy Director for Plans for Scientific Matters, a position that put him in charge of the technical aspects of covert and clandestine operations. Soon enough, Gottlieb would be launched into just such an operation himself.

Nineteen-sixty was a banner year for Africa: sixteen nations, including the Belgian Congo, were scheduled to become independent from their colonial masters. This was good news for the peoples involved, but it was not such a blessing in the eyes of U.S. State Department officials, who saw the nations as a potential new voting bloc that could tilt the balance of power away from the free West and toward the Soviet Union. The Congo, with its copper and diamond mines, was the largest and richest of the sixteen states scheduled for independence, and the feeling in the State Department was that as the Congo went, so would follow the rest of the newly emerging African nations. It was crucial, therefore, that the Congo not align itself with the Soviets.

In May 1960, the Congo held general elections. Candidates had made all sorts of wild promises during their campaigns, claiming that untold miracles would follow upon liberation. One advised voters to plant stones in the ground because they'd turn into gold on Independence Day. Another promised his constituents that if elected he would resurrect all their dead relatives. Patrice Lumumba, a former postal clerk who tailored his views to his audiences and persuaded conservatives, socialists, and communists alike that he would advance their respective agendas once in office, was elected prime minister.

The Congo was formally separated from Belgium on June 30, 1960. In the first week after independence, some 40,000 Belgians, many of them civil servants, fled the country. Public services broke down, government employees stopped getting paid, and by July 15—two weeks into the new era—chaos broke out in the Congo. To contain the damage, Lumumba applied to the Soviets for military aid.

It was around this time that the CIA's Deputy Director for Plans, Richard Bissell, had a couple of informal talks with his scientific adviser, Sid

Gottlieb, concerning the subject of the covert assassination of foreign leaders. Gottlieb suggested that biological agents were perfect for the task: they were invisible, untraceable, and, if intelligently selected and delivered, not even liable to create a suspicion of foul play. The target would get sick and die exactly as if he'd been attacked by a natural outbreak of an endemic disease. Plenty of lethal or incapacitating germs were out there and available, Gottlieb told Bissell, and they were easily accessible to the CIA.

This was entirely acceptable to Bissell. After all, he had already authorized a similar plan earlier in the year when, at the request of the CIA's Near East Division, he let Gottlieb mail a poisoned handkerchief to General Abdul Karim Kassem of Iraq, in an attempt to make him sick for a long time. Neither Gottlieb nor Bissell had ever learned, unfortunately, whether the handkerchief had reached the target; Kassem, at any rate, was later executed by a Baghdad firing squad, making the question moot.

In August 1960, CIA director Dulles sent a cable to Lawrence Devlin, the CIA station chief in Léopoldville, capital of the Congo, informing him that Lumumba's "removal must be a prime objective."

Gottlieb, meanwhile, reviewed a list of the biological agents to which he had access in the SO Division inventory at Fort Detrick. He wanted one that would either kill an individual or incapacitate him so severely that he would be out of action for the foreseeable future. Since the least suspicion would be aroused by an illness that was already common in the Congo, he narrowed down the list to seven or eight infectious diseases indigenous to the area, including tularemia, brucellosis, tuberculosis, anthrax, smallpox, and Venezuelan equine encephalitis. He decided on botulinum toxin and obtained a quantity of the agent from the SO Division.

He placed it in a jar together with enough chlorine to deactivate the toxin in case of emergency; the agent and the chlorine were physically separate but arranged in such a way that they could be mechanically mixed together if necessary. He then packaged the bottle so that it would pass for something else. Finally, he prepared a packet of accessory materials: rubber gloves and gauze face masks to be used for handling the stuff safely and a set of hypodermic needles for delivery of the agent to the target.

On September 19, Richard Bissell sent a cable to Lawrence Devlin in the Congo saying that a covert agent from headquarters would soon arrive in Léopoldville.

SHOULD ARRIVE APPROX 27 SEPT . . . WILL ANNOUNCE HIMSELF AS "JOE
FROM PARIS" . . . IT URGENT YOU SHOULD SEE "JOE" SOONEST POSSIBLE
AFTER HE PHONES YOU. HE WILL FULLY IDENTIFY HIMSELF AND EXPLAIN
HIS ASSIGNMENT TO YOU.

"Joe from Paris" turned out to be Sidney Gottlieb himself, who arrived
in Léopoldville on September 26. His biological assassination kit got there
separately by diplomatic pouch, the normal CIA conduit for sending such
mechanisms and devices.

Gottlieb now met with Devlin, told him of the assassination plans, and
instructed him in the use of the materials. The toxin had to be extracted
from the jar by means of the hypodermic syringe, he said, and then injected
into something that Lumumba would put in his mouth: food, water, tooth-
paste, or the like.

All of this was new to Devlin. He hadn't joined the CIA to assassinate
people, he thought it was a wild scheme, and he was skeptical that it would
work. Nevertheless, he was willing to go along with it, for on September 24
he had received a cable from Allen Dulles himself, saying:

WE WISH GIVE EVERY POSSIBLE SUPPORT IN ELIMINATING LUMUMBA
FROM ANY POSSIBILITY RESUMING GOVERNMENTAL POSITION.

Additionally, Gottlieb told Devlin that the operation had been approved
not only by Dulles but by "the highest authority," which was to say, the
president of the United States, Dwight D. Eisenhower.

Patrice Lumumba, however, was by this time in the protective custody
of the peacekeeping troops that the United Nations had sent to the area. He
was living in the former Belgian governor-general's residence in the Gombe
section of Léopoldville, a house set on a cliff high above the Congo River
where he was surrounded by armed guards.

Neither Devlin nor Gottlieb could penetrate the compound himself, nor
could a third operative whom Devlin hired for the mission. The botulinum
toxin, meanwhile, had been sitting in a drawer of Devlin's safe, unrefriger-
ated, for the past ten days. The substance was unstable, and its potency
had by now declined to the point that it could no longer reliably do its job.
Gottlieb took the jar from Devlin's safe, mixed the agent with the chlorine,
and threw the contents into the Congo River at Stanley Pool. Then he left
for Washington.

Further planning focused on more conventional methods. On October 15, Devlin received a cable from CIA headquarters suggesting:

POSSIBILITY USE COMMANDO TYPE GROUP FOR ABDUCTION LUMUMBA VIA ASSAULT ON HOUSE UP CLIFF FROM RIVER

to which Devlin replied, on October 17:

RECOMMEND HQS POUCH SOONEST HIGH POWERED FOREIGN MAKE RIFLE WITH TELESCOPIC SCOPE AND SILENCER.

All these theatrics were, in the end, quite unnecessary, for on January 17, 1961, Lumumba was removed from protective custody by the local Congolese, flown to Elisabethville in Katanga Province, and beaten to death by Katanga officials and Belgian mercenaries.

Three days later, on January 20, 1961, John F. Kennedy took the oath of office and a new presidential administration came to power in Washington. Shortly thereafter, Kennedy's secretary of defense, Robert S. McNamara, authorized a total review of the U.S. military picture. The review was a mammoth undertaking parceled out into some 150 separate study projects, and one of them, Project 112, concerned biological and chemical warfare. In his mission statement to the Joint Chiefs of Staff, who were responsible for carrying out Project 112, McNamara directed the investigators to "consider all possible applications, including use as an alternative to nuclear weapons. Prepare a plan for the development of an adequate biological and chemical deterrent capability, to include cost estimates, and appraisal of domestic and international political consequences."

The Joint Chiefs convened a working committee that included the four branches of service—the Army, Navy, Air Force, and Marine Corps—all of whom agreed that despite the work that had already been done on biological weapons over the years, such armaments were combat ready and operational in only the most limited sense. Nevertheless, biological and chemical weapons offered many advantages to the modern military, chief among which was their unique ability to produce "controlled temporary incapacitation" instead of the mass destruction and death caused by conventional high explosives and nuclear devices. The committee members therefore recommended that the nation enter on both a short-term crash program in biological agent manufacture and munitions production and a five-year program of research, testing, and development of new chemical and biological weapons systems.

Within a year of Kennedy's having assumed the presidency, then, the United States was embarked on a major new biological weapons program. The new program would be different from the old one in a number of ways. First, the biological agents would be disseminated by spray nozzles rather than by bombs, for the Army was by now finally convinced that spray dissemination of microbes was a far more efficient means of delivering pathogen to the enemy than bombs that killed the majority of the agent in the process of detonation. Second, to maximize the peculiar gift of biological weapons for producing controlled temporary incapacitation (CTI), the program would focus on incapacitating rather than lethal agents. Third, since all branches of the military were interested in biological devices of various sorts, all four services, and not just the Army alone, would participate in the planning and conduct of biological weapons tests, and in the evaluation of the test results. The four services, in addition, would each pay their fair share of the expenses involved.

Fourth and finally, nearly all of the new biological weapons trials would take place not within United States borders, as before, but rather at what the military referred to as "extracontinental test sites," the reason being that the new trials would be conducted on so large a scale that even Dugway Proving Ground, big as it was, would be too small to contain the test agents safely. A related consideration was that in order to be realistic, the test sites had to duplicate the respective climates of the likely target areas, which in the early 1960s were the Soviet Union and Vietnam. By and large, Russia was dry and arctic while Vietnam was wet and tropical, but neither of those two climate types was regularly found at Dugway. The proper atmospheric conditions existed in Alaska and in the central and south Pacific, and they were the main areas, therefore, in which the new biological weapons trials would have to be carried out.

By the end of 1961, Secretary McNamara had approved both the short-term crash production program and the five-year plan for an ambitious new series of chemical and biological weapons trials at extracontinental test sites. This being the sixties, however, McNamara and others within the defense community were concerned not only about the human health and

safety consequences of the upcoming tests, but also about the possible
adverse effects of large-scale open-air chemical and biological testing on
the environment. The prospect of spraying infectious agents across ten,
twenty, or thirty miles of open ocean gave rise to disturbing visions of
seabirds being killed by the hundreds or thousands, and even worse visions
of those birds carrying communicable diseases out of the test area and to
human population centers where they could proceed to infect, incapacitate, or kill people. Some of the seabirds of the central Pacific—the alba-
tross and the lesser and great frigate birds, for example—were known to
travel thousands of miles, and a wayward frigate bird had once turned up
in Maine.

Over the course of about a year, military planners at the Pentagon and
elsewhere developed a set of protocols and controlling procedures govern-
ing the location and conduct of the large-scale trials. Essentially, three
main requirements would have to be satisfied in order for a test to proceed.
One, a given biological agent could not be released into an area where it
was not already endemic: for example, the Army could not spread Q fever
on an island where it did not already exist in one or more animal species.
Two, a microbe should not be disseminated in an area whose bird population made it likely that the organism would be carried to regions inhabited
by humans. Three, the president himself would have to give prior approval
to any weapons trials that were likely to have protracted effects on the
physical or biological environment.

The Army could not escape the force of these regulations by performing
the trials at sea away from land areas because the central Pacific was frequented by pelagic birds—birds that fed, and in effect lived, on the open
ocean. It might be possible to find waters comparatively free of pelagic
birds, but the scientific knowledge of the distribution patterns and migration habits of the Pacific pelagic bird populations was too spotty for the
Army to pinpoint with any degree of accuracy exactly where those bird-
free areas might be located.

To conduct its upcoming biological warfare trials in such a manner as
would satisfy the Defense Department environmental protocols, therefore,
the Army would first have to discover what animal life and diseases
existed, particularly among birds, on both the open waters and the uninhabited tropical islands of the central and south Pacific. The Army, in other
words, would have to make a complete and systematic survey of the entire

bird population in an area comprising some 4.3 million square miles of open ocean.

The U.S. Army was not overrun by ornithologists, however, which meant that it would have to contract the work out to a group of competent professionals. Where in the country could you find a bunch of ornithologists who would undertake the task on behalf of the military, and especially in support of its biological warfare testing program? The answer was simple: the Smithsonian Institution.

Initially, it was to be called the Pacific Ocean Ornithological Project. Soon, though, someone realized what the acronym for that would be, whereupon the name was changed to the Pacific Ocean Biological Survey Program.

The project began in the fall of 1962, when officers representing the three principal military services, the Army, Navy, and Air Force, paid a visit to Remington Kellogg, the Smithsonian's Assistant Secretary for Science, at the institution's headquarters on the Mall in Washington. The officers were from the Deseret Test Center, a new military installation located in Salt Lake City, Utah, and they explained that for an upcoming series of biological warfare trials that they were planning in the Pacific, they wanted an ecological survey made of the entire tropical Pacific region. The Army scientists wanted to understand bird migration patterns and island ecologies in the area so that they could determine which islands, or what portions of the open ocean, would be safe for the large-scale testing of biological and chemical warfare agents. The Deseret Test Center's scientists also wished to know what diseases were endemic to the flora and fauna of the various islands so that they could be sure not to introduce any new pathogens to the region. The Navy and Air Force, for their part, wanted to know where certain bird populations were concentrated because sooty terns had a bad habit of choking up the jet engine intakes of aircraft stationed on some of these remote island bases.

All this sounded reasonable enough to Remington Kellogg. The fact that the Smithsonian Institution, according to its original mandate and charter, existed "for the increase and diffusion of knowledge among men" was no real obstacle to its participating in a secret program. Unavoidably, the Smithsonian's connection with the Army, and above all with the Deseret Test Center's biological warfare program, would have to remain secret, the officers said. Except for that prohibition, though, everything was open, free, and publishable. Whatever the Smithsonian's ornithologists discovered about the birds of the Pacific, whatever they learned about the flora and fauna of the islands, all those discoveries could be published in the open literature without any restriction. The knowledge generated by the survey, in other words, could be distributed via the normal scientific channels; it was only the Army's motives in funding the research that would have to be kept secret.

Kellogg discussed the project with some of the staff members who would be involved, and they all felt that the project would be wholly appropriate for the Smithsonian. It would be a source of needed funding, and it would lead to the increase and diffusion of knowledge among men. If that was the final effect of the survey, what did it matter that the Army was behind it?

Philip S. Humphrey, curator of the Smithsonian's Division of Birds, would be the principal investigator of the Pacific Ocean Biological Survey Program. In the next few months he hired some sixty-five people to go out and do the fieldwork. One of the first was a twenty-four-year-old ornithologist by the name of Roger B. Clapp.

Roger Clapp had just graduated from Cornell with a degree in vertebrate zoology and was looking for his first job when he heard through the grapevine that the Smithsonian was looking for zoologists for a massive Pacific island survey project. He sent a letter to Philip Humphrey, and Humphrey invited him in for an interview. Four months later, on October 2, 1963, Roger Clapp and four other Smithsonian researchers left Honolulu aboard a 400-ton, 107-foot-long U.S. Army tugboat bound for a first landfall at Howland Island.

Howland Island was barely two miles long and a half mile across at its widest point and had a total land area of only eight-tenths of a square mile. It was the least little speck of land in the Pacific, almost exactly on the equator. Yet during the latter part of the nineteenth century, Howland was the site of a flourishing guano trade, and approximately 125,000 tons of the stuff had been removed from the island. Howland had been colonized in the 1930s, and three separate runways had been built, each 150 feet wide, taking up seven percent of the land area. It was on account of the island's runways that when Amelia Earhart took off on her round-the-world flight in 1937, she intended to use Howland as a refueling point. She never reached the island, however, and was lost somewhere in the vicinity. The island was now an American possession and totally uninhabited.

Roger Clapp and his group on the tugboat had been out of sight of land for six straight days when at 8:58 in the morning of the seventh day they first laid eyes on Howland Island, a long low stretch of green absolutely out in nowhere. For Roger B. Clapp, just seeing it was an amazing thrill.

He and the others came ashore on a rubber raft and saw that the island was covered with birds. The birds were so tame that you could walk right up to them, and some of them actually fluttered up and landed on Clapp's shoulders.

They set up camp and had lunch: C rations, which in Clapp's case turned out to be turkey loaf with meatballs and beans, all of which he hated. Then they started collecting birds for blood: they caught ten masked boobies, four red-footed boobies, three red-tailed tropic birds, and one lesser frigate bird. One by one Clapp stuck a long-needle syringe through the breast and into the heart of each bird and withdrew enough blood to fill a small vial. All the birds survived the immediate puncture wound, but one of the masked boobies died shortly afterward, perhaps from shock, while Clapp was still holding it. Later he collected six vials of insects: thirty ladybugs, plus some sow bugs, spiders, beetles, and the like. He put both the blood samples and the insects on ice, and later transferred them to a refrigerator aboard the ship; at the end of the trip the samples would be flown to Fort Detrick, where the scientists would examine them for pathogens. Meanwhile, Larry Huber, the herpetologist, collected lizards, while Fred Sibley set out mouse and rat traps.

That afternoon Clapp walked the island making an inventory of all the masked booby nests and their contents, to get an overall population count. He saw the remains of the runway that Amelia Earhart was to have landed on, a long sandy strip parallel to the beach, discernible now only by means of the lighter vegetation. He saw that from any point on the island the day beacon was visible; it was a fifteen-foot-tall structure that looked like a lighthouse, but with a cylindrical red concrete top where the light would normally be. On the front of the day beacon was a weathered and chipped concrete plaque that said:

EARHART LIGHT
1937

That night, when the birds couldn't see the men approaching, Clapp and Sibley went around catching and banding masked boobies. The boobies were more than two feet long from beak to tail, and the procedure was to grab one with your hands, stuff it between your legs, and then in a quick motion reach down and slip the band onto one of the bird's legs before letting it go. The maneuver demanded a measure of finesse because a booby's bills were like intersecting steak knives, two long hard mandibles with serrated edges, and the birds had an exceptionally fast strike.

Roger Clapp banded eighty masked boobies his first night on the island. He would have done many more (his personal best would be 565 masked boobies in one night) if it hadn't been for the tidal wave false alarm. Near midnight, while he and Fred Sibley were happily grabbing and banding masked boobies, the tugboat, which had been riding quietly at anchor all this time, suddenly erupted like a Roman candle.

Lights!

Sirens!

Horns!

This scared the hell out of the island's birds, who all took off in a puff.

Clapp got on the radiotelephone to hear the tug's captain say: "A tidal wave's coming! We don't have time to come in and get you! Get to high ground!"

The maximum height of Howland Island was all of fifteen feet, with no coconut trees or anything else. Their only hope was the day beacon. The five men raced to it, climbed up the side, and tied themselves to the top with their belts. Two hours later, when it was clear that the threat, if any, had passed, they climbed back down again and resumed work. The captain, it turned out, was somewhat inexperienced in the area and didn't know when to ignore the tidal wave forecasts he heard on the radio.

The group spent five days on Howland Island. They then returned to the tug and sailed for Baker Island, an even smaller American possession a few hours away. It had been discovered in the 1800s by an American whaler, one Obed Starbuck, who claimed the place and named it New Nantucket. Someone had once abandoned a donkey there, and the island was so long and low that from a distance the donkey had appeared to be walking across the surface of the ocean.

The landing on Baker was somewhat risky—the island was surrounded by a coral reef and the surf was tremendous. There were more shipwrecks around Baker than practically anywhere else in the central Pacific. But they got in safely and repeated the whole survey process, counting and banding birds, taking blood samples, insect samples, and all the rest, for the greater good of science, not to mention the American biological warfare testing program.

And then, finally, they landed on Birnie. Birnie was one of the smallest islands in the Pacific, just sixty-four acres, part of the Phoenix Islands, which also included Enderbury, Sidney, Gardner, and Hull, all of which they ended up visiting. Birnie, however, was distinguished by the fact that, so far as anyone knew, it had never yet been explored by a trained scientist. Roger B. Clapp was thus a member of the first scientific team ever to set foot on the place.

In the end, Birnie proved to be not all that interesting. It was comparatively barren, with little wildlife in evidence: some boobies, blueberry noddies, black noddies, very little vegetation. There were also rats, which had eaten the tops off every last succulent plant they found on the island. They counted and sampled Birnie's entire bird, animal, and insect population in just two days.

Still, seeing the place with his own eyes, an unexplored tropical island of the Pacific, was the experience of a lifetime. For a young zoologist fresh out of college, it did not get any better than this.

14

hus far, the American germ warfare program had been a sober business, pursued with all the solemnity of purpose appropriate to a technology whose primary aim lay in making people sick or dead. Even the CIA's scheme for assassinating Patrice Lumumba with botulinum was a halfway serious plan that had a chance of working. But all sense of proportion took to the hills the moment the Kennedy administration set its sights on Fidel Castro.

The plots to get Castro, admittedly, went back to the closing days of the Eisenhower administration, and not all of them were assassination attempts. Starting in early 1960, the final year of Eisenhower's presidency, the CIA was already entertaining sundry plots for undermining Castro's sway over the Cuban population. His appeal, after all, was largely a matter of personal charm and charisma. His aptitude for giving mesmerizing speeches, his beard, even the ever-present cigar in his mouth, were primary factors in the overall equation. Therefore, each of them was targeted for destruction with one or another biological or chemical agent.

Someone in the CIA's Technical Services Division, for example, proposed the notion of destroying Castro's public image by spraying his broadcasting studio with a hallucinogenic drug similar to LSD. Supposedly, the drug would so badly undercut his speechifying powers that the dictator's hold over the Cubans would evaporate. But the chemical's effects were unpredictable, and so the CIA rejected the plan.

Then there was the scheme to make his beard fall out. This would be accomplished by dusting Castro's shoes with thallium salts, a depilatory, when he placed his shoes outside his hotel room door to be shined during the course of an upcoming trip. Faced with the horrifying prospect of a beardless leader, the Cubans would rise up in revolt en masse. The CIA's Technical Services Division went so far as to test the depilatory on animals, and apparently it worked. Castro, however, canceled the trip.

Then there was the poisoned cigar plot. A box of his favorite cigars would be spiked with so much botulinum toxin that he'd fall over dead shortly after lighting up. On August 16, 1960, a CIA agent obtained a box of the cigars in question and delivered them to the CIA's Office of Medical Services, where they were inoculated with the toxin and pronounced ready for use.

But there was no way to guarantee that Castro, and he alone, would ever smoke them. What if he started handing them out to a group of visiting schoolteachers? Much better to dispense the toxin in the form of a pill that could be dropped into Castro's drink: if done correctly no one would gulp it down but him, and meanwhile the assassin would have a chance to escape. Technical Services accordingly developed a batch of botulinum pills. The first prototypes unfortunately would not dissolve in water, but a second batch did, and in fact they worked so well when tested on monkeys, killing them all, that the chief of the CIA's Operational Support Division, James O'Connell, who was in charge of such things, immediately approved them for use against the Cuban dictator.

All that remained was the matter of getting them to Havana and into Castro's drinking glass. In light of their past failures in delivering lethal agents to within miles of Patrice Lumumba and other targets, the CIA had judiciously decided to seek the aid of competent professionals, which was to say, prominent and experienced underworld figures. To dilute the connection between such misbegotten souls and themselves, the CIA needed a go-between (a "cutout" in CIA nomenclature), and this it found in the person of Robert A. Maheu, a former FBI agent who was now a private detective in the employ of the recluse millionaire Howard Hughes. Soon Maheu had rounded up Johnny Rosselli, a midlevel Mafia figure active in the extortion business; Santos Trafficante, the Cosa Nostra chieftain of Cuba; and an unnamed Cuban exile who had an accomplice in Castro's favorite restaurant.

In March 1961, two months into the Kennedy administration, the four of them met outside the Boom Boom Room, a bar in the Fontainebleau Hotel in Miami. All of them went up to Maheu's hotel suite, where Maheu opened his briefcase and produced a large amount of cash plus the botulinum pills. He rolled them out in his hand: five or six gelatin capsules filled with a liquid. The pills, he told the Cuban exile, had to be placed in a cold liquid, not in coffee or hot soup, because heat inactivated the poison inside them. Also, the toxin wouldn't last forever, so the job had to be done soon.

But this plan, too, elaborate as it was, came to naught because Castro stopped eating at the restaurant where the accomplice worked. The Cuban exile later returned both the cash and the pills.

Four consecutive schemes and not one of them had made it to first base against Castro. By the start of 1963, the CIA had organized a separate section devoted to undercover operations in Cuba, the so-called Task Force W. Early that year Desmond FitzGerald, chief of the task force, proposed the next miracle device, an exploding seashell. Castro, he knew, was fond of skin diving. Why not rig up an exotic mollusk so that when he swam by to investigate, the thing would blow up in his face?

Sam Halpern, the executive officer of Task Force W, pointed out a few problems with this. For one, how could you be sure that Castro would be the one to trigger the device? Someone else, or maybe even a shark, could swim by, exploding the shell without accomplishing the objective. For another, the scheme left something to be desired as far as covert operations were concerned, inasmuch as seashells did not normally explode unaided. Castro blowing sky-high from the ocean floor would be a smoking gun, as it were, pointed directly at the United States.

But that idea, harebrained as it was, gave birth to another: the contaminated diving suit. Why not give Castro a wet suit whose inside was coated with an ugly fungus and whose breathing apparatus was contaminated with tuberculosis bacteria? It would be the dictator's last swim.

Richard Helms, who was then the director of the covert operations group (and later the director of the CIA), characterized this scheme as "cockeyed." Nonetheless, an agent from the Technical Services Division went out and purchased the diving equipment, dusted it with the proper microbes, and a short time later had it ready for shipment to Havana.

Castro's skin was saved on this occasion when an American lawyer (who was unaware of the plan) coincidentally gave Castro a different wet suit as a gift.

There was a final bionic device waiting in the wings, a poison pen. This was a ballpoint pen rigged with a hypodermic needle filled with Blackleaf 40, the trade name for nicotine sulfate, a tobacco extract commonly used as an insecticide but which was also highly poisonous to humans. Castro, in the process of signing some important document of state, would unwittingly inject himself with the toxin and a few moments later drop over dead.

During the fall of 1963, the CIA's Technical Services Division fabricated the poison pen, tested it, and found that it worked.

On November 22, 1963, in Paris, Desmond FitzGerald, the leader of Task Force W, offered the device to the Cuban dissident he'd hired to kill Castro and to whom he'd given the code name AM/LASH.

But AM/LASH, whose real name was Rolando Cubela, was not particularly impressed with the poison pen device. "Surely you can come up with something more sophisticated than that!" he said.

Almost as he spoke those words, President John F. Kennedy was himself assassinated in Dallas. According to the 1975 U.S. Senate report on assassination plots against foreign leaders, "It is likely that at the very moment President Kennedy was shot, a CIA officer was meeting with a Cuban agent in Paris and giving him an assassination device for use against Castro."

The poison pen, anyway, was the last of the CIA's exotic anti-Castro devices. Later schemes focused on hardware of the more conventional sort: high-powered rifles, hand grenades, and bombs. They, at least, could be depended on to work.

By the mid-1960s some of the CIA's considerable lust for gadgetry and intrigue had rubbed off on their brethren at Fort Detrick's SO (Special Operations) Division, whose secret agents now staged mock biological raids at the Greyhound bus terminal and at National Airport in Washington, D.C. The two operations were milestones of government paranoia and redundancy, for they only demonstrated yet once again what everyone knew well enough already, that concealed atomizers could fill up large open spaces with aerosol sprays.

One day in May 1965, nevertheless, a group of SO Division covert agents armed with spray generators hidden in briefcases duly filed into the Greyhound bus terminal in Washington, D.C., and set the machines going at five different points inside the main waiting room. Streams of dried *Bacillus globigii*, the anthrax simulant, wafted into the air. Other operatives, each equipped with a perforated suitcase containing a vacuum pump, took air samples at various points within the building: in the main lobby, at a door near the loading platform, in the restaurant, and in the game room, while a fifth secret agent wandered through the main waiting area and collected a "roving sample" of the atmosphere.

Within a few days the team staged essentially the same mock attack in the north terminal building of Washington's National Airport.

A year later they were at it again, this time in the tunnels of the New York City subway system, in an operation that would prove to be the high-water mark of the SO Division's covert attacks upon American public places. The SO Division's secret agents had two new inventions with them, the germ-filled lightbulb and the "Mighty Mite" air sampler, and the men were justly proud of both devices.

The lightbulb had come out of the SO Division's Commodity Development and Engineering Laboratory. It was an ordinary lightbulb with a tiny hole drilled through the base to allow the insertion of powdered simulant. Once the bulb was filled with the agent, the hole was sealed over, yielding a potentially lethal but utterly innocuous-looking device. What could be less suspicious, after all, than a lightbulb? The Mighty Mite air sampler was a motorized vacuum device that drew ambient air through a stack of filter paper. The apparatus, artfully concealed inside a trim little plastic case, was almost noiseless.

Between the 6th and 10th of June, 1966, the SO Division secret agents fanned out through three different lines of the New York City subway system, and from the platforms between the cars of moving express trains they dropped the filled lightbulbs onto the tracks. The lightbulbs shattered on impact, and the trains stirred up the air and aerosolized the agent, then dragged the cloud for a distance behind them.

Afterward, men with suitcase samplers collected air specimens at various points throughout the system. The official results, from the Army's report on the project, were not surprising: "Dropping an agent device onto the subway roadbed from a rapidly moving train proved an easy and effective method for the covert contamination of portions of subway lines."

What was surprising about the operation was the fatherly pride that the SO Division men took in the Mighty Mite. "I was an air sampler operator stationed in subway stations during the five tests of the New York Subway system in June 1966," wrote Special Operations Agent John C. Malinowski in his report. "I used a Mighty Mite Air Sampler, a Humidity-Temperature meter, and in one test I monitored incoming and outgoing trains every few minutes. During all operations I had a Mighty Mite Air Sampler at my side."

"While riding train to 23rd Street Station, a man asked me where I got the nice little plastic case (the Mighty Mite)," wrote another Detrick secret agent. "I told him all the hardware stores over town had them. He is going to buy one."

To supervise the large-scale extracontinental trials that were a necessary part of Robert McNamara's Project 112, the Army decided that it needed a new biological and chemical warfare headquarters located somewhere in the western United States. Fort Detrick, valuable as it had been in the past, was far too distant from where the new action was going to be, which was in Alaska and the Pacific. In May 1962, therefore, the Army moved into Buildings 103 and 105 at Fort Douglas, Utah, a disused Army base located on a bluff just east of Salt Lake City, and created an organization called the Deseret Test Center.

Despite its name, the Deseret Test Center would conduct no actual testing at Fort Douglas. The buildings housed only offices, and there were no laboratories, exposure chambers, pathogens, bombs, or hardware of any other type on the premises. The place was populated by an in-house staff of some 200 people and acted solely as a test planning center and information clearinghouse. Deseret was funded jointly by the Army, Navy, Air Force, and Marine Corps, all of whom could submit detailed requirements for weapons tests. Its operations bore such a high level of priority within the military that only combat priority exceeded it, and at one point the Deseret Test Center had eleven ships assigned to it from the Navy, including five tugboats and two large utility vessels called YAGs, the naval designation for "miscellaneous service craft."

Deseret's extracontinental tests were run as "safaris," the notion being that a small band of experts drawn from all over the military would gather together their gear and weapons and go off into the tropics in search of big biological game. There would be many such safaris over the Deseret Test Center's eleven years of existence, and the man in charge of them all was a Utah Mormon by the name of J. Clifton Spendlove.

Spendlove was a civilian scientist whose personal experience with chemical weapons went back to March 1945, when the U.S. Army laid down a smokescreen over the Rhine near Wesel, Germany, preparatory to an amphibious assault on German positions across the river. The smoke hid the American force well enough until a rift in the cloud allowed the Germans to fire a shell that landed some fifteen to twenty feet in front of Spendlove. The shell exploded, lifted him off the ground, blew him rearward some thirty feet, and flung shrapnel fragments into his right thigh. He regained consciousness eight hours later while being airlifted out of Germany.

Back in the States he acquired a Ph.D. in industrial microbiology from Ohio State and went to work at Dugway Proving Ground doing test design and analysis. During field trials with hot agents it was not unusual for him to spend six to eight hours straight, and in rare cases twelve hours straight, in a rubber suit and gas mask with no provision for eating, drinking, or any other bodily function other than breathing, which was itself a painful experience because you continually had to suck air in through the gas mask canisters. A day in that getup left your stomach sore. When he came to Deseret as test director—his formal title was Technical Director of the Plans and Evaluation Directorate—it was like being on vacation, especially when he was in the process of writing scripts for the documentary films that he produced, directed, and occasionally starred in.

Spendlove felt that the time was not far off when public opinion and world politics would turn against biological warfare altogether—pathogen production, weapons testing, munitions stockpiling, everything—and so to record the tests for history, he decided to make a film of each one. The films were factual and straightforward but the shoots had required substantial amounts of advance planning, scripting, and editing nonetheless. The camera angles had to be right, the on-screen action had to be clear and intelligible, and so on, and Spendlove was responsible for every last facet of the final production.

There was the time they went to Alaska, for example, to do some tularemia tests in the Tanana River valley. He went up there with Paul Adams, Fred Houston, and some others from Deseret, bringing with them an ample supply of rhesus monkeys. Alaska was the stand-in for Russia, and the area was so remote that they'd had to carry rifles to protect themselves and the monkeys from bears, and so cold that they'd had to heat the nozzles of the E-2 spray generators to keep them from freezing up (but not heat them too much because excess heat killed the pathogen). Even so, when the aerosol sprayed out it often formed an ice fog that suspended itself in stratified layers over the river. It was so cold up there, in fact, that they'd had to put the monkeys in tiny parkas in order to keep them warm, charming little zip-up bags with hoods on top so that only their faces would be exposed to the air. Each monkey had its own individual Deseret Test Center biological-field-trial parka, and all of this was lovingly captured on film.

Spendlove's films were so realistic and convincing that periodically he'd put together a sequence of short excerpts from the year's tests and show them at the annual planning conference, where they functioned as advertisements for further testing and continued funding. The military bigwigs in the audience loved them, and the Deseret Test Center never lacked for either work or funds.

Over its eleven years of existence, the Deseret Test Center planned, conducted, and evaluated the results of about a hundred separate chemical and biological weapons trials, each with its own official project name or number, at various extracontinental test sites around the world. The project titles ran the gamut from AUTUMN GOLD, a ship-penetration study, to YELLOW LEAF, a chemical weapons test in the Panama Canal Zone, plus many more in between. There were BIG EYE, BIG JACK, and BIG TOM; there were EAGER BELLE, ELK HUNT, FEARLESS JOHNNY, FLOWER DRUM, GREEN MIST, HALF NOTE, NIGHT TRAIN, PURPLE SAGE, RED OAK, SUNDOWN, and WHISTLEDOWN, among a slew of others. Deseret's first major biological trial was so long in the preparation that it had two successive project names: RED BEVA and SHADY GROVE.

While the project was still in the early formative stages the name RED BEVA seemed innocuous enough: all it meant was *R*esearch and *D*evelopment, *B*iological *Eva*luation. The idea was to go out on safari into the tropical Pacific, the stand-in for Vietnam, and spray a line of hot agent (UL or Q or both) over a stand of monkeys on the deck of a boat. The planners wanted to use a pleasure craft to disguise the true nature of the operation, and so in the fall of 1963 or the spring of 1964 a man from the Deseret Test Center, passing himself off as Barney Franks, a member in good standing of the moneyed classes, waded into the southern California private boat market and purchased *Freedom II*, an eighty-foot-long wooden-hulled yacht with twin diesel engines that was currently berthed in San Diego Bay.

First it would have to be converted into a proper biological warfare test vessel—but secretly, as all of this was even more hush-hush than usual. And so in a separate project (Project Piggyback), the Army set up a fictitious export-import firm in San Diego's Old Town to take delivery of a succession of autoclaves, centrifuges, and the like, and install them aboard the *Freedom II*. All this was done, and within a short time two Navy officers sailed the boat to Hawaii for a shakedown cruise. But with all the new hardware and equipment on board, plus fuel oil sloshing around in fifty-five-gallon drums on the deck, the craft was not entirely seaworthy, and the test designers ended up abandoning the ship itself, scuttling the craft off Hawaii, and, with it, the concept of a private germ warfare test vessel. They also got rid of the name RED BEVA, which was now deemed far too intelligible for comfort.

Out of RED BEVA's ashes came SHADY GROVE, the title being contributed by an official in the Pentagon whose function it was to dream them up and who periodically attended performances at the Shady Grove Dinner Theater in suburban Maryland. SHADY GROVE was more ambitious than RED BEVA in every way. The object now was to demonstrate both high-speed, low-altitude spray dissemination as well as large-area coverage of hot agent, and so the plan was to go out into the Pacific and have an Air Force or Marine Corps jet fighter lay down a long line of agent over Deseret's small naval fleet and see how far they could get the pathogen to travel before it lost effectiveness. If and when the test ever happened, it would be the largest biological warfare trial ever conducted.

SHADY GROVE, indeed, would far exceed in scope and size the second wave of sea trials conducted by the British. After the end of Operation HARNESS in February 1949, John Dudley Morton had come back to Porton Down and, together with David Henderson, hatched a new scheme for dispersing hot agents over stands of animals on the open water. Instead of using a trotline of dinghies they'd line up the animals on a barge and douse them all with hot agent from a point upwind. The barge would make for a stable working environment, and the crew members could move about it in relative comfort and safety. The whole affair would be *under control*, something Morton never thought of the landing ship and the rubber dinghies as being.

The Royal Navy offered him a surplus spud pontoon, a large rectangular steel float that had first seen use in the Normandy landings and that had later been converted to a ship-repair platform off the coast of Malta in the

Mediterranean. Although it was essentially nothing more than a floating steel box with watertight compartments belowdecks, to John Dudley Morton and the rest of the test crew, the pontoon's vast flat surface was like their own miniature aircraft carrier.

In 1951 the Navy towed the pontoon to protected waters in the English Channel off the Isle of Wight, and Morton and his men ran simulant aerosols over it, first from the deck itself and then from a dinghy towed by a launch. The scheme worked so well that the naval shipyard at Chatham added an airtight control house with filtered air supply, a place where the crew could take refuge during hot agent trials. With that improvement the pontoon became the stage for four more years of open-air biological trials at sea: Operation CAULDRON in 1952 and Operation HESPERUS in 1953, both of them off the west coast of Scotland. Then in 1954 and again in 1955, the *Ben Lomond* towed the pontoon 4,000 miles across the Atlantic for the final two British tests, Operations OZONE and NEGATION, both of which took place in the Bahamas. As always, Americans from Camp Detrick were there observing and assisting in every case. When they finally undertook their own sea trials, therefore, the Americans had a large store of experience to draw on.

SHADY GROVE took place in 1964 on the open waters off Johnston Atoll, some 800 miles southwest of Honolulu. The Smithsonian Institution's scientists had visited the area earlier and found that it was not in fact the ideal site for a massive release of pathogens: the region's frigate birds had turned up thousands of miles to the west, as far away as China and Russia. Three countervailing factors proved irresistible to the Deseret Test Center's planners, however. First, the principal agent to be disseminated (UL, *Pasteurella tularensis*) was already present in the area's flies and mosquitoes, so that additional amounts deposited during large-scale trials would have no appreciable effect on the geographical extent of the disease. Second, UL was known to be an exceptionally fragile and feeble microorganism, and any new microbes deposited on bird feathers would likely be killed by ultraviolet rays within a few hours of sunrise. Third, Johnston Atoll had a long paved runway suitable for jet aircraft, and, most important of all, the entire place already belonged to the U.S. Army Chemical Corps, who used it as a chemical weapons storage center. At one point, Johnston Atoll held a stockpile of two million pounds of the nerve agent sarin ("GB") on a total land area of approximately one square mile.

The Deseret Test Center's naval fleet went under its own code name, Project SHAD, an acronym that stood for "Shipboard Hazards and Defenses."

The vessels included five light tugboats (LT-2080, LT-2081, LT-2085, LT-2086, and LT-2087) and two YAGs (YAG-39 USS *George Eastman* and YAG-40 USS *Granville S. Hall*). These were the very ships that had hauled the Smithsonian's bird project scientists all over the Pacific, a function they still performed whenever they were not participating in Deseret's biological or chemical warfare trials.

All the ships in the Project SHAD fleet were "citadel" ships, meaning that for hot agent trials all of their portholes and hatches could be shut tight, and the intake air passed through HEPA filters—high-efficiency particle arresters. Citadel ships also had sophisticated washdown systems that got rid of any stray agent after the tests. The tugboats, in addition, were equipped with a "doghouse" at the aft end—a small, sealed, air-filtered hut with modest lab facilities for collecting and analyzing air samples.

The two YAG vessels were converted Liberty ships. YAG-40, the *Granville S. Hall*, was the laboratory ship, the equivalent of the British *Ben Lomond*. It had been fitted out with machinery and equipment that included gimbaled lab tables to ensure stability on the open ocean. Cliff Spendlove, the test supervisor, would often spend two or three weeks aboard *Granny Hall*, directing test operations and documentary filming. Both of the YAGs had helicopter landing pads on the bow so that personnel, equipment, or animals could be transferred between ships in a precisely choreographed aerial ballet, during which the two helicopters took off simultaneously, passed in midocean, and traded places on the opposite ships.

Over a number of weeks in February and March 1964, Project SHAD's five light tugs stationed themselves at distances that varied from one to twenty miles apart in a long line off Johnston Atoll. When the tugs were at maximum distance from each other they formed a sampling line that was fully 100 miles long.

The biological agent would be laid down by a Marine Corps A4D Skyhawk, a small and trim attack bomber that had a wingspan of less than that of a Piper Cub and a top speed of 670 mph. A squadron of the A4Ds was stationed on the Johnston Island air base.

The trials began shortly after sunset. During the day, a couple of WV-2 reconnaissance aircraft, Lockheed Constellation radar planes known as "Willie Victors," swept the downwind area to warn away any fishing boats, pleasure craft, or other ships that might be in the agent cloud's line of travel. Then, at night, when the winds died down and the weather conditions favored the air inversion that the scientists looked for because it held the agent close to the water, one or more A4D Skyhawks equipped with spray tanks, disseminator nozzles, and a supply of the biological agent took off from the Johnston Atoll airstrip and headed for the line of tugs.

The first release was a line of FP, fluorescent particles. The target ships were lined up parallel with the wind, and the plane flew across the top of the line as if crossing a T and sprayed the particles into the night air.

The plane roared by, turned, and then came back around for a second pass, this time releasing a line of BG, *Bacillus globigii*, the anthrax simulant.

On the target ships, the scientists collected cloud samples with mechanical sampling equipment. Over the maximum separation distance of 100 miles, it sometimes took eight hours for the cloud to travel down the entire line of ships.

In the final phases of the SHADY GROVE series, the plane released lines of hot agent over animal specimens. The two hot agents used were UL (*Pasteurella tularensis*) and OU (*Coxiella burnetii*, the Q fever microbe), and the animals were rhesus monkeys. The sailors had treated the monkeys well enough beforehand and had often played with them on deck. Occasionally a crew member would use a Magic Marker to tattoo a monkey's chest with hearts, flowers, or the name of a girlfriend.

All that changed immediately after exposure, when the monkeys had become hot objects. Lab crew members dressed in protective gear crated up the animals and high-lined them back to the *Granville S. Hall*, where they awaited the customary test-subject fate.

There were twenty trials in the SHADY GROVE test series, and in the biggest of them the plane sprayed a thirty-two-mile-long line of agent that traveled for more than sixty miles downwind before it lost its infectiousness, thereby establishing that a jet aircraft flying at or close to Mach 1 could lay down a line of pathogenic agents over an immense geographical area in an incredibly brief span of time.

In the summer of 1968, four years after SHADY GROVE, the Deseret Test Center's private biological fleet of tugs and the YAG-40, the USS *Granville S. Hall*, gave their final performances. The stage setting this time was Eniwetok Atoll, an irregular oval-shaped group of low coral islands that enclosed a broad lagoon about twenty-five miles in diameter. Eniwetok was part of the Marshall Islands in the central Pacific, some 2,000 miles to the west of Johnston Atoll.

Between 1948 and 1958, the United States had conducted forty-three nuclear weapons tests on the islands and in the Eniwetok Atoll lagoon. Then, after the last of them, the Americans had come in with earth-moving equipment and over a three-year period, at an expense of $120 million, had scraped up 111,000 cubic yards of radioactive soil and debris and removed it to the tiny island of Runit within the atoll, which itself had been the site of about a dozen of the nuclear blasts. The Americans buried the radioactive waste in a bomb well left by one of the explosions, a depression called Cactus Crater, measuring 30 feet deep and 350 feet across, and capped it off with a dome made of 358 panels of eighteen-inch-thick concrete. The rest of Eniwetok Atoll, as a result, was no longer radioactive.

On June 21, 1966, Dayle Husted of the Smithsonian's Pacific Ocean Biological Survey Program, had visited Eniwetok Atoll while aboard the U.S. Coast Guard cutter *Basswood*. He noted only a few bird species during the two days the cutter spent in the lagoon—brown noddies, white terns, and wedge-tailed shearwaters, among others—but he later compiled a list of some thirty-two bird species that had been observed on the atoll's islands over the years.

Despite the profusion of bird life on the atoll, the Deseret scientists chose Eniwetok for their final major biological weapons trial because the area had a tropic marine environment similar to that found in the Panama Canal Zone, where Deseret also had a test center, and in Vietnam. In addition, the area's ecology had been so disturbed already by the nuclear bomb tests that a few wee microbes weren't going to matter much. And of course the entire atoll was essentially a U.S. satrapy.

So in the summer of 1968, the scientists came in with their boats and monkeys, their space suits and jet fighter planes, and they sprayed clouds of "PG" (staphylococcal enterotoxin B, familiarly known as SEB), an incapacitating biological agent, over the area. Deseret, by this point, had given up the practice of naming its tests, and this one would be known only by a number: DTC Test 68-50. Leader of the safari was Edgar ("Bud") Larson, the Detrick scientist who in the early 1950s had detonated the first hot agent bomb inside the 8-Ball.

Other than for the facts that the animals were placed on some of the atoll islands as well as on the boats, that much of the sampling equipment was mounted on 300-foot-high towers on the atoll islands, and that the planes were now McDonnell-Douglas F4 Phantom jets piloted by U.S. Air Force Vietnam War veterans, the Eniwetok trials proceeded largely as did those off Johnston. The Air Force also had a new dissemination tank they wanted to test, the so-called AB45Y-4, a long, thin projectile that resembled a V-2 rocket.

An unclassified paragraph in the Deseret Test Center's final report on DTC Test 68-50, a two-volume, 244-page work written by John H. Morrison, told the entire story in six terse sentences: "Test 68-50 was a full-scale field test of the area coverage potential of the F4/AB45Y-4 incapacitant weapon system. The weapon system disseminated the aerosol over a 40–50 km downwind grid, encompassing a segment of the atoll and an array of five tugs. Stability of the bulk agent and of the agent aerosol was evaluated by the response of animals to the intravenous injection of graded doses. The agent proved to be stable and did not deteriorate during storage, aerosolization, or downwind travel. A single weapon was calculated to have covered 2400 square km, producing 30 percent casualties for a susceptible population under the test conditions. No insurmountable problems were encountered in production-to-target sequence."

Twenty-four hundred square kilometers was equal to 926.5 square miles, an area roughly twice the size of metropolitan Los Angeles.

At the end of DTC Test 68-50, the Deseret Test Center's final series of large-scale extracontinental field trials, there was not the least shred of doubt in anyone's mind that biological warfare was a workable technology. In terms of large-area coverage, no other weapons system even came close.

15

and then, just as the building was complete, the foundation collapsed. Late in 1969, a year after the Deseret Test Center's highly successful trials at Eniwetok Atoll, Richard M. Nixon, president of the United States, summarily and unilaterally terminated the American biological warfare program.

"I have decided that the United States of America will renounce the use of any form of deadly biological weapons that either kill or incapacitate," he said on November 25 in a statement issued by the White House press secretary. "I have ordered the Defense Department to make recommendations about the disposal of existing stocks of bacteriological weapons. Our bacteriological programs in the future will be confined to research in biological defense, on techniques of immunization, and on measures of controlling and preventing the spread of disease."

As to why he was suddenly canceling the twenty-five-year-old germ warfare project at the very apex of its experimental success and operational readiness, Nixon's explanation was that "biological weapons have massive, unpredictable, and potentially uncontrollable consequences. They may produce global epidemics and impair the health of future generations. Mankind already carries in its hands too many of the seeds of its own destruction. By the examples that we set today, we hope to contribute to an atmosphere of peace and understanding between all nations."

That, anyway, was the rhetoric offered up for public consumption. It did not find many backers at Fort Detrick, where the claim that biological weapons were "unpredictable" was viewed as especially galling. Some of the Detrick researchers had spent half their lives collecting data about the behavior of biological agents in a wide variety of different conditions and environments. They could tell you about the percentage of live agent that would survive explosive or spray dissemination, about the cloud's rate of travel downwind, its lateral dispersion patterns, its physical and biological decay rates under various temperature and humidity conditions, night or day. They could tell you about the concentration of viable pathogen per unit of elapsed time, the total number of likely casualties in various population settings, and a million other details.

Unforeseen changes in the wind and weather could throw a given prediction off, but every statement about the future was subject to some degree of uncertainty. Predictions about the effects of biological munitions were no worse than most, and were valid and accurate within their known limits.

The claim that biological warfare might "produce global epidemics" was likewise viewed as sheer fantasy, at least insofar as American biological weapons were concerned. The fact was that the only agents the U.S. Army had ever weaponized were those that had extremely low, and in some cases nonexistent, rates of person-to-person transmission and secondary spread. Anthrax, the granddaddy of all biological warfare agents, had never been observed to travel from person to person. Brucellosis was rarely, if ever, transmitted person-to-person, and so on down the list of America's biological warfare agents. Portraying them as producing "global epidemics" was just silly, the Detrick crew thought, as was the equally ludicrous claim that biological weapons might "impair the health of future generations." Clearly, this was politics speaking, not science.

Detrick's biological warriors therefore began to speculate about what had been Nixon's actual motivation. There had been some public demonstrations at Fort Detrick, but such activities were nothing new at weapons centers. A group calling for an end to biological weapons research had picketed Detrick in 1961 and had planted a "Peace Tree" outside the fence. But the tree died and the research continued.

More recently, in the summer of 1969, a new wave of demonstrators showed up at the gates to protest *chemical* weapons research at Fort

Detrick. But Detrick had never done chemical work, and Richard Nixon, in any case, was not notably sensitive to demonstrations, rallies, peace marches, or picketing. Suspicion therefore centered on the Dugway sheep-kill incident.

According to Dugway Proving Ground's incident log, the event had begun with a telephone call on Sunday, March 17, 1968:

> At approximately 1230 hours, Dr. Bode, University of Utah, Director of Ecological and Epidemiological contract with Dugway Proving Ground (DPG), called Dr. Keith Smart, Chief, Ecology and Epidemiology Branch, DPG, at his home in Salt Lake City and informed him that Mr. Alvin Hatch, general manager for the Anschute Land and Livestock Company, had called to report that they had 3,000 sheep dead in the Skull Valley area.

Skull Valley was a thirty-mile-long stretch of open desert north of Dugway, and several ranchers ran sheep in the area, including members of the Gosiute Indian Reservation, which bordered the proving ground. For 3,000 sheep to die virtually overnight, there seemed to be only one logical explanation: one of Dugway's chemical or biological agents had escaped the test area, passed over the sheep, and killed them. The case was bolstered when the Army admitted to having released a quantity of VX nerve agent over a Dugway test site the previous Wednesday, March 13. On that occasion the shut-off valve on the plane's spray tank failed to close and the aircraft continued to release the stuff during a pull-up to higher altitude, where the winds aloft apparently carried the chemical far beyond the intended target range.

Plausible as it was, that scenario was undercut by the fact that no other livestock in the area, and in fact no other animals of any type, including cows, horses, dogs, rabbits, or birds, appeared to have suffered any ill effects, a circumstance that was hard to explain if VX had in fact caused the sheep deaths. An alternative explanation put forward by some in the Dugway/Deseret community (including Cliff Spendlove) was that the deaths were due to an organic phosphate insecticide that the ranch owner had allegedly sprayed within a mile of the sheep just the day before they had started dying. Whatever the cause, the Army ended up paying out $1 million in damages to the sheep owners while nevertheless refusing to concede that it was responsible for the deaths.

In any case, the Dugway sheep-kill incident did not logically account for why Nixon had canceled the *biological* program, which was a thing apart from the chemical works. The alternative explanation was that the president's motive was political, an attempt to send a message to other nations that biological warfare was really not worth the trouble.

Detrick veterans took a dim view of that tactic. Abandoning the offensive program, they thought, was a dangerous precedent, for the real message it would send was that the country was unable to respond in kind to a biological attack, which was in effect an engraved invitation for other nations to develop their own biological weapons systems and direct them against the United States.

Still, in his original statement Nixon had spoken only of "biological" and "bacteriological" weapons, and had said nothing whatsoever about toxins—agents such as botulinum and SEB, both of which the Americans had weaponized and stockpiled. The Detrick scientists never knew whether this omission was intentional or accidental, but they nevertheless viewed it as a loophole and a means by which they could continue on with their work. All they had to do was to rewrite their research proposals so as to focus on toxins instead of bacteria, viruses, and rickettsiae.

But that plan of action, so logical and satisfying at the time, did not succeed, for on February 14, 1970, Valentine's Day, Nixon formally extended his biological weapons ban to include toxic agents as well.

The entire offensive program, then, was finally over and done with.

By the time Nixon terminated it, the U.S. Army had officially standardized and weaponized two lethal biological agents, *Bacillus anthracis* and *Francisella tularensis*, and three incapacitating biological agents, *Brucella suis*, *Coxiella burnetii*, and Venezuelan equine encephalitis virus (VEE). The Army had also weaponized one lethal toxin, botulinum, and one incapacitating toxin, staphylococcal enterotoxin B (SEB). Supplies of these agents had been mass-produced and were currently stockpiled at the Army's Directorate of Biological Operations at Pine Bluff Arsenal.

The arsenal, which occupied 15,000 acres eight miles north of Pine Bluff, Arkansas, had been built in 1941 as a chemical weapons repository. In 1953 the Army added on a biological weapons production, filling, and storage complex. The production plant, Building 50, held a total of eighteen fermenters, ten of 3,600-gallon capacity, three of 5,000-gallon size, and five

7,000-gallon fermenter tanks. With a combined capacity of 86,000 gallons, the fermenter building at Pine Bluff was smaller than the old Vigo plant in Indiana, which had a total fermenter capacity of 240,000 gallons, but even so it could turn out a substantial quantity of agent.

When Nixon canceled the program, Pine Bluff Arsenal had already produced vast quantities of the standardized microbes and toxins, had filled bombs, bomblets, spray tanks, and assorted other munitions with them, and had stockpiled the filled weapons in underground igloos, refrigerated vans, and the other storage facilities at the base. That aggregate weapons cache was the forbidden fruit of twenty-six years of research and experimentation. Now, suddenly, all of it would have to go.

Supposedly, according to the first estimates, it wouldn't take long to get rid of it all. Most of the agents were highly perishable, and after the judicious application of heat, light, and disinfectant, the microbes and poisons would be dead and gone. Existing stocks could be destroyed, said the Defense Department, "well within a year."

Still, this was the first time that anyone had been faced with the task of systematically going through vast amounts of stockpiled pathogen and disposing of it once and for all, safely. There would be an element of risk involved at every stage, and hanging over the entire process was the specter of a biological accident in which a store of virulent agent would be unintentionally released on the public.

There was the separate problem of ferreting out every last bit of infectious agent—every bomb and bomblet, spray nozzle, flask, test tube, and vial of the bacterial, viral, and rickettsial microorganisms that researchers had cultured in large-scale fermenters, and in smaller laboratory quantities, at Detrick, Pine Bluff, and elsewhere over the last twenty-six years. The worry was that some kindly and gentle researcher would want to save out, as a souvenir of his quarter century's worth of hard work in the lab, a tiny relic of his pet pathogen, a share of which he'd one day bring home and put in the deep freeze for posterity.

And then of course there was the CIA, whose private stock of microbes and poisons was kept under lock and key in Building 1412, a malign-looking structure with more smokestacks than windows that was the new home of Detrick's SO Division. The CIA was a world unto itself and it was anyone's guess how they'd respond—if at all—to Nixon's destruction order.

So the task of "demilitarization," as it was called, actually reduced to three separate problems: one, making sure that all offensive agent stocks were actually located; two, destroying those stocks except for authorized quantities held out for legitimate (defensive) research purposes; and three, proving that the materials were truly dead and gone and could pose no threat to anyone ever again—the problem of "verification of destruction."

This being the United States, and the disposal process being a military operation, every conceivable government regulatory, oversight, environmental protection, or other bureaucratic unit jumped on the agent-destruction bandwagon, including the Department of Health, Education, and Welfare; the Department of the Interior; the Department of Agriculture; the Environmental Protection Agency; the President's Council on Environmental Quality; the Armed Services Explosives Safety Board; the Air Force Armament Laboratory; and the Centers for Disease Control, plus the full roll call of state and local public health officials. With the combined intervention of all those agencies, a task that was scheduled to be finished "well within a year" had not even been started more than a year after Nixon's Valentine's Day announcement. In the end, the entire destruction, disposal, decontamination, and certification process would take more than three full years—and even then, some of the stocks would be overlooked, misplaced, or "lost."

The focal point of the action was the Directorate of Biological Operations at Pine Bluff, home of the vast bulk of the nation's offensive microbes and weaponry. The Army drew up a set of operating procedures and established a verification office on the site. The personnel would make an inventory of the materials to be destroyed, then they'd kill the pathogens, melt down the munitions, and send samples of whatever remained to the Centers for Disease Control, whose laboratory technicians would make their own independent assay of the materials. The whole operation was portrayed as a historic "swords into plowshares" event, with all the attendant press briefings, news releases, and live television coverage.

For the next three years, death came to the nation's supplies of offensive agents in the form of heat. Workers at Pine Bluff emptied bomb casings of their anthrax spores and heated the spores for three hours at 280°F. They chemically treated the sludge, heat-sterilized it a second time, and then plowed the final residue into the Pine Bluff ground soil while a public relations spokesperson explained what a good fertilizer it would make. Other workers smelted the bomb casings in furnaces at 2,000°F, cooled the slag, tested it for signs of pathogen, and then buried the remains in a sanitary landfill on the site. They did the same with storage drums, cans, containers, and packaging materials. They steam-sterilized the fermenters in Building 50 at 250°F for three hours, tested their insides for signs of life, and then cut up, crushed, and sold off the remainder as scrap metal.

The empty facilities of the Pine Bluff biological complex then underwent a systematic decontamination. Space-suited personnel came in and disassembled safety cabinets, bomb-filling equipment, and the associated ductwork and piping, washed it down with disinfectants, and then subjected it and the entire surrounding area to a treatment of formaldehyde gas.

On May 1, 1972, more than two years after Nixon's Valentine's Day statement, and after a total cleanup cost of $10,830,600, the Pine Bluff complex was declared fit for civil use and turned over to the Food and Drug Administration, who planned to house the National Center for Toxicological Research on the premises.

The Army applied similar decontamination measures at Fort Detrick, except that Building 470, the anthrax pilot plant, remained standing, with all the original fermenter tanks, plumbing, and everything else left intact. It was as if the Army had placed the facility on indefinite hold, just in case it should ever be needed again.

There had been some gallows humor in the pilot plant at the time of official decommissioning among those afraid of losing their jobs. Someone had scrawled the words "Neck-stretching room" on the door of room 604. Inside the room, a rope with a noose at the end hung from an overhead pipe, and a sign on the wall said: "The rope. If you must use the rope please leave a quarter on the tank. Purpose is to clean your blood up."

On Monday, February 16, 1970, two days after Nixon's Valentine Day announcement, Nathan Gordon, chief of the CIA's Chemistry Branch, went in to see his boss, Sid Gottlieb, chief of the Technical Services Division, to find out what ought to be done with the stock of biological agents and toxins they had in storage at Fort Detrick.

Gottlieb and Gordon agreed between themselves that unless they rescued the stuff soon, the Army would eventually destroy it along with everything else. That would leave the CIA unable to mount the very type of covert operation for which the biological agents had been stockpiled in the first place. Those biological supplies had been procured over a number of years at great cost, and it would be a shame to just throw them all down the drain. Besides, the Nixon order seemed to be aimed at the military; the CIA clearly was not part of the Department of Defense, and so technically the destruction order might not apply to them.

Gordon suggested to Gottlieb, in view of all this, that they might want to ask the authorities at Detrick to put aside the CIA's own private stocks and withhold them from destruction. Neither of the two wanted to make such a request on their own authority; any such recommendation would have to come from higher up, from the current director of the CIA, Richard Helms. So Gottlieb asked Gordon to prepare a memorandum for Helms listing the CIA's agents and toxins at Detrick, and suggesting a plan of action to save them. The memorandum, which bore the subject line "Contingency Plan for Stockpile of Biological Warfare Agents," revealed that the CIA had no small biological treasure chest at its beck and call:

Under an established agreement with the Department of the Army, the CIA has a limited quantity of biological agents and toxins stored and maintained by the SO Division at Ft. Detrick. The agents and toxins are:

Agents:

1. Bacillus anthracis (anthrax)—100 grams
2. Pasteurella tularensis (tularemia)—20 grams
3. Venezuelan Equine Encephalomyelitis virus (encephalitis)—20 grams
4. Coccidioides immitis (valley fever)—20 grams
5. Brucella suis (brucellosis)—2 to 3 grams
6. Brucella melitensis (brucellosis)—2 to 3 grams
7. Mycobacterium tuberculosis (tuberculosis)—3 grams
8. Salmonella typhimurium (food poisoning)—10 grams
9. Salmonella typhimurium (chlorine resistant) (food poisoning)—3 grams
10. Variola virus (smallpox)—50 grams

Toxins:

1. Staphylococcal Enterotoxin (food poisoning)—10 grams
2. Clostridium botulinum Type A (lethal food poisoning)—5 grams
3. Paralytic Shellfish Poison—5.193 grams

4. Bungarus Candidas Venom (Krait) (lethal snake venom)—
   2 grams
5. Microcystis aeruginosa toxin (intestinal flu)—25 mg
6. Toxiferine (paralytic effect)—100 mg

This stockpile capability plus some research effort in delivery systems is funded at $75,000 per annum.

In the event the decision is made by the Department of Defense to dispose of existing stocks of biological weapons, it is possible that the CIA's stockpile, even though in R&D quantities and unlisted, will be destroyed.

If the Director wishes to continue this special capability, it is recommended that if the above DOD decision is made, the existing agency stockpile at SO Division, Ft. Detrick be transferred to the Huntingdon Research Center, Becton-Dickinson Company, Baltimore, Maryland. Arrangements have been made for this contingency and reassurances have been given by the potential contractor to store and maintain the agency's stockpile at a cost no greater than $75,000 per annum.

A day or two later, however, Sid Gottlieb had changed his mind about sending the memo to Helms. Instead, Gottlieb and Gordon had a personal meeting with Helms and Thomas H. Karamessines, the Deputy Director for Plans, all four of whom now jointly decided that the CIA would comply with Nixon's destruction order fully and in all respects. They assigned Gordon to so inform the Detrick authorities, and shortly thereafter he drove up to Frederick and met with the post commander, Colonel E. M. Gershater, and Andy Cowan, chief of the Special Operations Division, telling them that they were free to dispose of the CIA's private stockpile as they wished.

That, thought Gordon, was the end of the matter. But soon after he'd returned to his office at CIA headquarters at Langley, Virginia, Gordon got a call from Andy Cowan at Detrick concerning the 5.193 grams of paralytic shellfish poison (code-named "SS" and also known as saxitoxin) that was part of the CIA inventory. Didn't the CIA want the stuff? It was theirs, after all. They had paid for it.

The shellfish toxin, Gordon knew, fell into a somewhat special category. It was a strong poison, a single gram of which was enough to kill 5,000 people. The CIA had used it to make suicide pills (agency people called them "L-pills"), and Francis Gary Powers, the U-2 pilot, had been carrying one (and was supposed to have swallowed it, but didn't) when his plane had been shot down over Russia in 1960. The toxin was an extremely rare commodity that had had to be laboriously extracted from mussels, clams, and other shellfish, and Edward Schantz, the Detrick microbiologist, had once gone up to Alaska and collected thousands of butter clams in order to isolate a small amount of it. The five grams in storage at Detrick in fact represented fully fifteen percent of all the saxitoxin that had ever been isolated anywhere in the world. The CIA had indeed paid the Army good money for it—$194,000, which had been disbursed to the two U.S. Public Health Service labs that had isolated the toxin.

Acting on his own, and without informing anyone higher up in the chain of command, Nathan Gordon told Cowan that he'd take the shellfish toxin off his hands if someone at Detrick would physically transport the stuff to the CIA warehouse in Washington.

A day or so later SO Division agent Walter Pannier entered room 202 of Building 1412 at Detrick, unlocked safe B172C3, and withdrew from it two 1-gallon cans. Each had a piece of paper stuck to the top. They said:

PARALYTIC SHELLFISH POISON
WORKING FUND INVESTIGATIONS
U.S.P.H.S., Taft Center, Cincinnati, Ohio
5.000 grams

And:

PARALYTIC SHELLFISH POISON
WORKING FUND INVESTIGATIONS
Northeast Shellfish Sanitation Center, U.S.P.H.S.
Narragansett, R.I.
5.927 grams

The two cans therefore held a combined total of almost 11 grams of shellfish toxin, enough to kill 55,000 people.

In late February 1970, Walter Pannier and Charles Senseny, also of the SO Division, drove the cans of shellfish toxin to the CIA biological storage facility at the U.S. Navy Bureau of Medicine and Surgery on 23rd Street in northwest Washington. They put the cans in a small freezer tucked under a workbench in room B10, a basement storage area. There they remained, for the next five years.

On April 10, 1972, just as the cleanup of Pine Bluff Arsenal drew to a close, a new biological weapons convention was opened for signature in Washington, London, and Moscow. The document, titled "Convention on the Prohibition of the Development, Production and Stockpiling of Bacteriological (Biological) and Toxin Weapons, and on Their Destruction," was four pages long and consisted of fifteen separate articles, of which the first two were primary and controlling:

Article I

Each State Party to this Convention undertakes never in any circumstances to develop, produce, stockpile or otherwise acquire or retain:
(1) Microbial or other biological agents, or toxins whatever their origins or methods of production, of types and in quantities that have no justification for prophylactic, protective or other peaceful purposes;
(2) Weapons, equipment or means of delivery designed to use such agents or toxins for hostile purposes or in armed conflict.

Article II

Each State Party to this Convention undertakes to destroy, or to divert to peaceful purposes, as soon as possible but not later than nine months after the entry into force of this Convention, all agents, toxins, weapons, equipment and means of delivery specified in article I of the Convention, which are in its possession or under its jurisdiction or control. In implementing the provisions of this article all necessary safety precautions shall be observed to protect populations and the environment.

Representatives of seventy-nine nations, including the United States, the Soviet Union, Great Britain, and Canada, signed the biological weapons treaty immediately. Still, there was a major difference between signing a treaty and ratifying it, and the agreement was not binding on any signatory until and unless it was ratified by that nation's government, which in the case of the United States meant approval by the Senate and the president. The United States had signed the Geneva Protocol against bacteriological warfare in 1925, but the Senate had refused to ratify it, and so the protocol was never legally binding on the country. The United States had always abided by its terms, however, since the protocol prohibited only the *use* of biological weapons but neither the development nor the stockpiling of them, and the United States had never in fact used any of the biological weapons it had developed.

On December 16, 1974, finally, the U.S. Senate ratified the new biological weapons treaty. President Gerald R. Ford himself ratified it a month later, on January 22, 1975, and the treaty officially entered into force on March 26, 1975.

If all the signatory nations actually proceeded to abide by the terms of the document they had set their hands to and formally approved, it appeared that the world would finally be free of the worrisome specter of biological weapons, and that Nixon's strategy in abandoning the American germ warfare program would be vindicated after all.

Except that the Russians, despite their having both signed and ratified the 1972 biological weapons convention, had never abandoned their own germ warfare program. In fact, they had accelerated and expanded it.

This became clear to the old-line biological warriors at Fort Detrick the moment they heard reports of the 1979 anthrax outbreak in the Russian city of Sverdlovsk. The data on the outbreak were admittedly scant, and although the event had actually occurred in April 1979, the first news of it didn't reach the West until February 1980, when the German magazine *Bild Zeitung* published an account of a military installation in Sverdlovsk having accidentally released a cloud of anthrax spores that the prevailing winds carried into the city suburbs, causing an unknown number of deaths.

The Russian explanation that the anthrax came from contaminated livestock that had somehow slipped by the city's meat inspectors was implausible on a number of counts. For one thing, the Russian public health service was thought to be highly effective, and it was hard to imagine it failing in so gross a fashion. Second, satellite photography of the area showed that the installation in question, known as Military Compound 19, had all the ear-marks of a biological warfare research facility, with security fencing and armed guards, massive ventilation machines, and a huge complex of animal pens. Third, the symptoms of those who'd died—collapsing on the streets before they could even reach the hospital—sounded like inhalational anthrax and not the gastrointestinal variety of the disease. All the circum-stantial and medical evidence therefore pointed to a respiratory infection that could only have been delivered by aerosol—as in fact it had been.

Many years afterward, in the spring of 1992, Russian president Boris Yeltsin confessed that "our military developments were the cause" of the Sverdlovsk outbreak. Not only had the Russian government been in viola-tion of the 1972 Biological Weapons Convention at the time they signed it, they had consistently violated it ever since. Indeed, the nation had begun the largest biological weapons buildup in its history in 1973, the year after they'd signed the treaty, when the Soviet Central Committee and the Council of Ministers created a network of biological warfare research cen-ters collectively known as Biopreparat.

Biopreparat far exceeded, in size, scope, and ambition, anything the Americans had ever done in the field of germ warfare. At its height, the Rus-sian operation employed 25,000 people at eighteen research centers and six bacterial production plants across the country, plus a pathogen storage com-plex in Siberia and a field test site on Vozrozhdeniye ("Rebirth") Island in the Aral Sea—the Russian equivalent of Gruinard Island. Their researchers had worked with the same basic list of agents—anthrax, tularemia, and Q fever, among others—that the Americans had long experimented with, and they'd invented many of the same dispersal methods and devices, including a refrig-erated warhead containing a quantity of skittering munitions that fanned out in all directions and spread over a large area when released.

Beyond those similarities, however, the Russian and the American pro-grams differed fundamentally. The Russians had weaponized smallpox, an agent that the Americans had initially investigated but had soon abandoned. Smallpox was an especially nasty pathogen to place inside a biological war-head. Unlike the bacteria and viruses weaponized by the Americans, none of which was transmissible from person to person, smallpox *was* easily trans-missible from one person to the next, and was fatal in approximately ten percent of cases. An initial epidemic of the disease could therefore spread out in secondary, tertiary, and further waves of infection, and could travel widely outside the area of initial release.

But there was a second reason why smallpox was especially objectionable as a biological warfare agent, which was that the disease had been eradicated in 1977, after an intensive eleven-year-long worldwide medical effort. To disseminate smallpox from a biological weapon, therefore, would be to reintroduce into the world a disease that the public health services of several nations, including the Russians themselves, had successfully wiped off the face of the earth. The Soviet biological warriors, however, had massproduced and stored tons of dried smallpox virus in their weapons bunkers.

They had also experimented with Marburg virus, cause of an extremely rare African hemorrhagic fever that had a human mortality rate of approximately twenty-seven percent. The Russians did not confine themselves to existing microorganisms and diseases, however. Beginning in 1983 they used genetic engineering techniques in order to create a range of pathogens that were either more virulent than those found in nature or resistant to conventional Western antibiotics, or both—microbes that could cause "superplagues."

One was *Pasteurella tularensis*, the cause of tularemia. In 1983, at the Biopreparat research center at Obolensk, sixty miles south of Moscow, the Russian scientists developed a new strain that was more virulent than those previously known and more amenable to dissemination by aerosol. Then in 1985 they turned their attention to *Yersinia pestis*, cause of bubonic plague, the Black Death. Plague, like smallpox, had high rates of person-to-person transmission and secondary spread, but by 1987 the Russian scientists had reached the point where they could manufacture 200 kg of a "super-bubonic-plague" organism per week, enough to kill 500,000 people.

The Russians were not the sole violators of the biological weapons convention, however. In 1996 U.S. intelligence sources claimed that twice as many countries had or were actively pursuing the development of offensive biological weapons in 1996 as when the biological weapons convention came into force in 1975. Those nations included Libya, North Korea, Iraq, Taiwan, Syria, the Soviet Union, Israel, Iran, China, Egypt, Vietnam, Laos, Cuba, Bulgaria, and India, most of whom had both signed and ratified the treaty.

From all of this it was abundantly clear that if Richard Nixon's motive in terminating the American germ warfare project had been to provoke other nations into abandoning their own biological weapons programs, then his strategy was one of the grandest failures in recent political history.

■   ■   ■

The great mystery of biological warfare, in the end, was why it had never been used. Other than for small-scale sabotage attempts that amounted to live field experiments and isolated covert attacks against individuals, biological weapons had never been used by any nation, either in war or peace. This made them highly unusual: practically every other armament that had ever been invented, everything from the crossbow to the atom bomb, had been used at least once on the battlefield, including the chemical gases mustard, chlorine, and phosgene. Biological weapons seemed to be the lone exception.

Military historians had several explanations, the most common of which was that biological agents had not been used because they could come back and infect the aggressor—the "boomerang" effect that had been recognized as far back as 1932 by Leon Fox. But long-distance dissemination *had* no boomerang effect: the effect was completely escaped by high-altitude bombing, aircraft spray dissemination, biological warhead–equipped missiles, and the like. Many nations had possessed such weapons, but none had ever used them.

A second explanation was that biological agents were too dependent on the changes of wind and weather to be relied upon to accomplish a given combat mission. But chemical gases were equally dependent on atmospheric conditions, and they had nevertheless been used repeatedly in World War I by the Germans, British, and French, and even by the Americans.

A third explanation was that biological weapons were uniquely repugnant and morally objectionable, that there was something specially reprehensible about directing against one's fellow human beings the causative agents of those very pestilences that mankind had worked so hard to overcome throughout the ages.

But why? Why was it worse to die from a disease (which people did continually in the normal course of events) than from bullets, bombs, or nuclear radiation? Why was the use of a biological agent that produced a mild and temporary illness more objectionable than the use of a weapon of mass destruction that obliterated a city?

Indeed the case could be made that biological weapons were *less* morally objectionable than other types since they were in an obvious way far more "natural" than high explosives or nuclear bombs, neither of which existed in nature. Pathogens were component parts of planet Earth, items that contributed to, expanded, and enlarged the sum total of planet Earth's biodiversity.

Biological warfare was in effect "green" warfare, a method of killing people with the very agents that nature itself had created and regularly applied to the task. In the words of Al Webb, the Detrick biologist who'd assayed the Operation HARNESS results in Antigua, biological warfare was "a natural way of dying."

The true reason why biological warfare had never been waged had less to do with such tactical considerations as the boomerang effect, its unpredictability, or a visceral sense of its moral repugnance. The reason was that biological weapons lacked the single most important ingredient of any effective weapon, an immediate visual display of overwhelming power and brute strength.

The final goal of warfare, after all, was not to kill, maim, or sicken—those were only means to an end. The end of warfare was rather to force the adversary to surrender and submit, to knuckle under and yield to the pressure of greater strength. Such strength was shown by a violent display of physical force, a visual manifestation whose effects were immediately apparent and physically overpowering.

High explosives provided an obvious display of physical force. Nuclear weapons provided a show of force that approached the level of omnipotence. Biological weapons were just the opposite: they were silent, secret, invisible, and slow. They were excellent killing machines, but they made exceedingly poor weapons.

An effective weapon stunned, stupefied, and bullied the enemy into submission with a sudden manifestation of overwhelming power. Biological weapons did none of those things. And that was why no one had ever used them.

Epilogue

before they were entirely finished with the place, the U.S. Army Chemical Warfare Service found one more use for Horn Island in Mississippi. In June 1946, shortly after the SS *Francis L. Lee* docked at the U.S. Naval Magazine wharf at Theodore, Alabama, with a cargo of more than 3,000 tons of captured German bombs, rockets, and grenades, it was clear that some of the chemical munitions were leaking their store of liquid mustard agent. The Army entertained a number of plans to deal with the situation, the first of which was to dump all the leaking bombs in the Gulf of Mexico, and they did in the end throw thirty 1,000-pound bombs and three 500-pound bombs into the Gulf. As for the more than a hundred munitions that were still leaking, the Army went ahead with plan two, which was to place the bombs aboard a barge, haul them to Horn Island, and dispose of the mustard agent by burning. This was Operation Hornblow.

By mid-July the Army had transported the leaking bombs together with a heavy-duty mobile crane, twenty barrels of waste oil, and a small demolition crew to Horn Island. Dressed in full protective gear, the men test-burned one 500-pound bomb by placing it in a shallow trench, surrounding it with dry wood, pine needles, and pine cones, and dousing it all with waste oil. The kindling went up in a flash, whereupon a sharpshooter fired a volley of shots from a .30-caliber rifle, rupturing the bomb case. The liquid mustard agent blew out and ignited, burning so rapidly that it resembled a short burst from a flamethrower.

Over the next few days, the men set fire to the rest of the bombs, stacking them in piles on the beach on the north side of the island. When the last batch, a stack of forty-three 500-pound mustard bombs, exploded, some of the bomb cases were thrown laterally several hundred feet. A fireball rose high into the sky, a miniature atomic bomb–like cloud appeared overhead, and the sound of the blast was heard as far away as Pascagoula.

Then the men loaded the bomb cases—a few of which, despite everything, were still leaking mustard agent—back onto the barge, towed it out to open water between Horn and Petit Bois Islands, and heaved the cases over the side. Except for the faint odor of mustard detectable in the sand, the burn area itself was clean, and on August 3, 1946, the Army finally departed the place.

On June 15, 1947, the Army transferred ownership of Horn Island to the Department of the Interior, and the island later became part of Gulf Islands National Seashore. Every so often a contemporary visitor to Horn Island, which is still reachable only by boat, will come upon a railroad spike or a small glass vial on the white sands and wonder how it ever got to this remote island wilderness.

In 1947, an ad appeared in the *New York Times* for surplus government property near Terre Haute, Indiana. The property consisted of 700 acres of land, buildings housing 20,000-gallon fermenter tanks and process equipment, plus assorted other outbuildings, labs, and accoutrements. Pfizer Inc., the pharmaceutical company, took a lease on the Vigo plant, which they planned to convert to the production of antibiotics. The production system, they soon learned, was not in flawless shape: the pipes leaked and wouldn't hold pressure, sterility was poor, and there was a problem with effluent disposal. But the company made repairs and improvements, remodeled some buildings and got rid of others by pushing them into the lake at the edge of the campus, and finally started manufacturing drugs.

Pfizer Inc. owns and operates the Vigo plant to this day, and still produces veterinary-grade antibiotics in the Chemical Warfare Service's original 20,000-gallon fermenter tanks.

■ ▩ ▨

After the four successive investigating teams returned to Camp Detrick from Japan, the data, autopsy reports, slides, and other materials they had collected from Shiro Ishii and his subordinates were placed in four large trunks and made available to Detrick researchers, none of whom found the information to be all that helpful. The data in most of it, they found, were too crude to be of any use in predicting the behavior of a given weapon system. For example, the Japanese had sprayed anthrax spores on humans, but their reports gave no actual spore counts; instead, they gave the weight of the wet spore paste that was used in the spraying, which was not an item of useful information.

Ed Hill, the investigator who'd wanted the Japanese scientists to be "spared embarrassment" for their wartime activities, gave an in-house seminar on the material that he and Joe Victor had brought back to Detrick. The researchers attended his talk; they listened, asked questions, went away, and that was the end of it. By and large, the Detrick scientists trusted their own current research over secondhand information derived from a foreign program years ago, and so the interview transcripts, the thousands of specimen slides, and the autopsy reports (two of which were 350 pages long, and one of which was 800 pages long) proved to be material that nobody wanted. Detrick shipped it to Dugway for storage, Dugway sent it back to Detrick for final disposition, Detrick shipped it back to Dugway a second time for formal accession into their technical library, and the technical library finally got rid of it by sending the stuff to the Library of Congress, where it remained.

Arvo T. Thompson, who'd been Ira Baldwin's executive assistant, was transferred to U.S. Army Headquarters in Japan. Thompson had been the second Detrick interrogator to go to Tokyo, having been preceded only by his friend Murray Sanders.

Thompson was a trusting soul and had been widely liked at Detrick. By the time all the Japanese data had come back, however, it was clear that he'd been thoroughly gulled by Shiro Ishii during their interviews.

In 1950 Thompson was promoted to colonel and posted at Tokyo. He took a room at the Dai Iti Hotel, a large residential building frequented by American soldiers.

On Friday, May 18, 1951, at about 4 A.M., Arvo Thompson took his .45-caliber Army pistol out of its holster, pointed it at the center of his forehead, pulled the trigger, collapsed forward on top of the gun, and died.

Shiro Ishii, on the other hand, never stood trial for anything, and lived peacefully in Tokyo until his death from throat cancer on October 9, 1959, at the age of sixty-seven. A relic of his bones was buried beneath the Unit 731 memorial, the Seikon Tower, in Tokyo's Tama Cemetery.

Many of Ishii's former staff members at Ping Fan rose to become successful businessmen, civic leaders, and physicians and surgeons. Ryoichi Naito, who'd tried to get yellow fever virus samples from Rockefeller University in 1939, attempted to make a biological weapon out of fugu toxin at Ping Fan, and then acted as Murray Sanders's interpreter in Tokyo, later founded the Green Cross Corporation, makers of artificial blood. Before it went out of business, the company had branches in England and the United States.

Ishii and his men had dynamited the Ping Fan complex as the Russians were advancing toward it in the closing days of the war. Afterward, the Chinese restored the headquarters building and converted part of it into a high school and another part into a museum devoted to Ishii's wartime experiments.

At the end of Operation NEGATION in the Bahamas, John Dudley Morton came back to Porton Down and reported the results to a man by the name of Horton who worked for the British MI6 intelligence section. He showed up in Horton's office and announced to the receptionist: "Morton from Porton, to see Horton."

He always liked saying that line.

Try as it might, the United States could never successfully extricate itself from the charge that it had waged germ warfare in Korea. Simple denials had all the force of a sane person denying charges of being crazy.

"Categorically and unequivocally these charges are entirely false," said Secretary of State Dean Acheson on March 4, 1952. "The U.N. forces have not used, are not using, any sort of bacteriological warfare."

Which of course was exactly what he'd say no matter what.

On May 22, 1952, General Ridgway, the UN commander, said: "No element of the United Nations High Command has employed either germ or gas warfare at any time."

And so on.

The Americans got a measure of relief back at the United Nations, where in the Security Council on July 1, 1952, American representative Ernest A. Gross called the Communist charges a "big lie" and proposed that the UN order an official inquiry to determine once and for all whether there was any truth to the charges.

The proposal came up for a vote on July 3, with all votes in favor except for one against by the Soviet Union. The Soviet veto, according to the American representative, meant that "they know the charges cannot bear the light of day."

The captured American fliers were released from captivity after the war, at which time all of them recanted their extorted germ warfare "confessions."

Years later, in 1969, the UN secretary-general issued a report on chemical and biological weapons, which concluded that "there is no military experience of the use of bacteriological (biological) agents as weapons of war," and that "there is no clear evidence that these agents have ever been used as modern military weapons," all of which implied that the case for the American use of biological weapons in the Korean War was weak at best.

The real problem with the alleged evidence, however, was not that it was too weak but that it was too strong. There was far too much of it, and the evidence was lavish beyond belief. Indeed, if the Korean charges were true, then in 1951 and 1952 a tidal wave of pathogens, insects, plants, and animals had fallen out of the skies over North Korea and China in a manner that was so bold and obvious it was as if the planes were littering the countryside with smoking guns.

That scenario, however, did not fit the chief operational feature of the American germ warfare program, which was that from the start and forever afterward it had been conducted under strict and total military secrecy. The American government was paranoiac about the program, branding every last document CONFIDENTIAL, RESTRICTED, SECRET, or TOP SECRET. The authorities swore personnel to silence upon pain of fine, imprisonment, or both. They forbade researchers from talking about their projects to anyone outside their own small circle of coworkers. And the government enforced it all to the hilt, with a network of internal spies at Camp Detrick, Horn Island, Dugway, and elsewhere, who rigidly policed the code of silence.

The Army placed the secret installations behind barbed wire, in facilities disguised to hide their true function. They had shipments of needed supplies and equipment sent under bogus manifests to false addresses. They conducted open-air tests at some of the world's remotest locations, and often under the cover of night.

And then, suddenly, in the middle of all this camouflage, cover, secrecy, and concealment, U.S. fighter planes start dropping biological payloads out over a known war zone in broad daylight? Not just once or twice, quietly and experimentally, but repeatedly, frequently, regularly, to the point that, according to the final tally offered by the Chinese and North Koreans, American planes had made more than 1,165 open and obvious germ warfare attacks upon the Far East?

No.

Much later, in January 1998, Yasuo Naito, the Moscow-based reporter for the Japanese newspaper *Sankei Shimbun*, found twelve documents in the Russian Presidential Archive that showed how, beginning in 1952, a group of midlevel North Korean, Chinese, and Russian operatives had forged evidence to support the germ warfare allegations against the Americans. One of the documents, a memorandum from Soviet secret police chief Lavrenti P. Beria to Soviet premier Georgi Malenkov, described how "two false regions of infection were simulated for the purpose of accusing the Americans of using bacteriological weapons in Korea and China. Two Koreans who had been sentenced to death and were being held in a hut were infected. One of them was later poisoned." Another document described how the Koreans had obtained cholera bacteria from corpses in China and then used the organisms to simulate an epidemic. And so on.

In the fall of 1998, the *Bulletin* of the Cold War International History Project at the Woodrow Wilson Center in Washington, D.C., published an English-language translation of all twelve Russian documents, together with analyses and commentaries by two experts who decided, on the basis of them, that the germ warfare charges against the Americans had been fabricated out of whole cloth.

At the end of the offensive program, the Army turned part of Fort Detrick over to the National Cancer Institute, whose scientists pursued peaceful nondefense-related cancer research. For itself, the Army built a new medical institute: the U.S. Army Medical Research Institute of Infectious Diseases, abbreviated as USAMRIID. USAMRIID's mission was to develop vaccines and treatments for infectious diseases, especially those that other countries might use in biological warfare against the United States.

That was an activity that went back to the earliest days at Camp Detrick. Over the twenty-six years of its existence from 1943 to 1969, Detrick's scientists had developed twenty-seven vaccines against many of the diseases they had studied for offensive purposes, including anthrax, tularemia, brucellosis, plague, and Q fever. They had developed toxoids against five different types of botulism and skin tests for brucellosis, tularemia, tuberculosis, tetanus, and anthrax.

In addition, the Detrick researchers had devised containment strategies for working with hot agents, including both the safety cabinets and protective clothing later used at the National Institutes of Health, the Centers for Disease Control, and elsewhere. Beyond that, the air sampling techniques they had invented found peaceful uses in environmental air sampling, air pollution monitoring, and the like. From these and other spin-off applications, it was possible to conclude that the offensive biological warfare program was not in fact, as it had sometimes been portrayed, an unalloyed disgrace to humanity.

In 1975, the building that housed the 8-Ball burned down, but the sphere itself, which had survived hundreds of bomb explosions within, emerged from the flames intact. Two years later, in December 1977, the 8-Ball was placed on the National Register of Historic Places, but in normal Detrick fashion the identifying plaque did not say why, stating only: "This property has been placed on the National Register of Historic Places by the United States Department of the Interior."

Building 470, the anthrax pilot plant, underwent two successive decontaminations to remove all remaining traces of anthrax and the other infectious agents that had been produced there over the years. Supposedly the procedures succeeded and the building's 44,000 feet of usable space were fit for public occupancy. Nobody on the base moved in, however, except for the flocks of local pigeons who flew in through the broken windows on the upper floors and set up housekeeping in the cooling towers.

The place was a white elephant. Nobody wanted it and nobody knew what to do with it, and in 1988 the Army gave the building to the National Cancer Institute. They, too, did absolutely nothing with it or its 3,000-gallon anthrax fermenters, and the pilot plant still stands there today looking much as it ever did.

Not until 1975 did the Olson family learn that Frank Olson had been the victim of an LSD experiment just a week before his death. Even then, they learned it only from a *Washington Post* story about the Rockefeller Commission's investigation into illegal CIA activities. The *Post* account, which told how a man fitting Olson's description had jumped from a New York City hotel window after an unwitting dose of LSD, caused a minor sensation at the time. Within a year the Olson family was invited to the White House, where they received an official apology from President Gerald R. Ford and a settlement from the federal government in the amount of $750,000.

That, however, was far from the end of the Frank Olson story. On March 19, 1978, Olson's daughter Lisa, her husband Greg Hayward, and their two-year-old son Jonathan died in a plane crash while they were en route to upstate New York, where they were to visit a lumber mill they were planning to invest in with their share of the settlement money.

In 1994, upon the death of his mother, Olson's elder son, Eric, who had never accepted the theory that his father jumped out the window of his own accord, had his father's body exhumed from the grave in Frederick, Maryland, and examined by a team of pathologists. Removed from the coffin, the body was as black as a piece of charcoal, and the three consulting specialists could not agree among themselves whether the forty-one-year-old vestiges of Frank Olson's injuries were indicative of murder or were wholly explainable as the effects of a jump. In 1996, however, Eric Olson took the results to an assistant district attorney in New York City, Stephen Saracco, who investigated the case for more than two years without bringing any charges against Robert Lashbrook, who had been with Olson the night he died, or Sid Gottlieb, who had administered the LSD, or anyone else.

Gottlieb, meanwhile, had long since retired to "Blackwater," his country estate off Turkey Ridge Road, a narrow and winding dirt path outside the rural village of Boston, Virginia, where he lived with his wife, Margaret, and their little brown mongrel, Missy. The large modern house on the 50-acre property had solar heating, 6,000 square feet of living space, and a swimming pool. In 1998, now approaching old age, the Gottliebs moved a few miles away to a two-story redbrick house in the small town of Washington, Virginia, in Rappahannock County on the eastern flank of the Shenandoah Mountains.

The past was finally catching up to Sid Gottlieb, however, for in early 1999 he was facing trial in New York City for allegedly having slipped LSD (or some other mind-altering drug) into the drink not of Frank Olson but rather of one Stanley Milton Glickman, an American artist living in Paris.

From all appearances, the Glickman affair was a sort of a prequel to the Olson experiment. According to court documents, in October 1952, while in Paris, Gottlieb and some other Americans had fallen into a lengthy political debate with Glickman at a restaurant known as the Café Select. After the argument, "one of the men offered Glickman a drink as a conciliatory gesture, and Glickman eventually accepted. Rather than call over the waiter, the man walked to the bar to get the drink." After drinking half of it, Glickman "began to experience a lengthening of distance and a distortion of perception," while simultaneously "the faces of the gentlemen flushed with excitement."

The question to be decided in court was whether it was in fact Sid Gottlieb who had given the drink to Glickman. The case against Gottlieb was heavily circumstantial, Glickman had died in 1992, and there was no assurance that Gottlieb, for all his past escapades, would ever stand convicted of anything.

And then on March 7, 1999, two weeks before the Glickman trial was to begin, Sidney Gottlieb died. He had suffered congestive heart failure, contracted pneumonia, died at the University of Virginia hospital in Charlottesville, and was cremated.

Three weeks later, on a cold and gray Saturday afternoon, a "Memorial for Sid" was held in the auditorium of the old Rappahannock High School, an aging two-story white clapboard structure across the street from the Gottlieb house. A hundred or so people attended, including former CIA agents, friends and relatives, and members of the local hospice where Sid, a lifelong stutterer, did speech therapy work with stroke victims.

A train of people came to the open microphone and sang the praises of Sid Gottlieb who, in retirement, had written poetry, been a member of a Zen spiritual group, and acted in the annual Shepherd's play at Christmas. Finally a youngish man in a rain parka stepped to the microphone and said: "Anyone who knew Sid knew he was haunted by something." The man wanted the audience to join him in a prayer to bury this thing, whatever it was, so that Margaret and the rest of the Gottlieb family could have peace.

There followed a moment of silence.

In April 1975, as part of a search of its holdings here and there around Washington, D.C., the CIA discovered two cans of shellfish toxin in a room at a storage facility at the U.S. Navy Bureau of Medicine and Surgery. They had been there, untouched, since Walter Pannier and Charles Senseny had deposited them in the freezer some five years earlier. William E. Colby, then director of Central Intelligence, notified the White House of this find, which in turn led to a major U.S. Senate investigation of "Unauthorized Storage of Toxic Agents" in September 1975. (The CIA investigators had also unearthed eleven milligrams of cobra venom.)

The toxins were once again slated for destruction, which was supposed to take place at Edgewood Arsenal on June 11, 1975. But at the last minute, on June 10, the CIA decided not to go ahead with the plan after all, as they "wished to consider further ways of insuring that the destruction of the material could not later be misinterpreted." Having escaped death yet once more, the toxins remained in the safe in which they'd been found.

Later, at the end of the Senate hearings into the unauthorized storage of the shellfish toxins, the senators themselves proved unwilling to destroy the stuff, for on Thursday, September 18, 1975, they drafted and formally approved a letter to CIA director Colby telling him: "If adequate safety and security cautions could be taken, and if it is consistent with our treaty obligations, the Committee believes that it might be appropriate for the CIA to consider donating these toxins to properly supervised research facilities which can use these poisons for benign uses such as curing such debilitating diseases as multiple sclerosis."

Sometime thereafter, the CIA gave the toxins back to Edward Schantz, the Detrick microbiologist who had isolated much of the original stocks, and who had since returned to private academic research at the University of Wisconsin. Schantz parceled out the toxins to private researchers, who used them for peaceful and benevolent purposes.

In late April 1998, thirty former members of Project SHAD, the cover name for the convoy of ships involved in the Smithsonian's Pacific Ocean

Biological Survey Program and the Deseret Test Center's field trials near Johnston Island and Eniwetok Atoll, held a reunion at the Holiday Inn San Diego Bayside, in San Diego, California.

Most of these men had not seen each other in thirty years, but soon enough it was like old times. They toured San Diego one day, another day they went into Tijuana, and the final night they were back on a ship for a harbor cruise with dinner, dancing, and booze.

Several of the old-timers brought pictures from the days aboard the light tugs or the *Granny Hall* or the *George Eastman*. Others brought memorabilia—old menus, coffee cups, and such—and all had their own tall tales to tell from the time they steamed across the Pacific in tugboats that were never meant to leave the harbor, and had even towed one of them for a distance of 865 miles across the ocean, watching as the landlubber scientists from the Smithsonian got seasick from the ship's rolling and pitching only a modest 30 degrees.

On August 20, 1998, at his home on South Regency Circle in Tucson, Arizona, Ira Baldwin, lucid and, except for considerable hearing loss, in generally good health, celebrated his birthday. He was 103.

Three weeks later, on September 12, 1998, Cliff Spendlove, the former technical director of the plans and evaluation group at the Deseret Test Center, held a reunion of Deseret old-timers at his A-frame summerhouse in Weber Canyon, near the High Uintas Wilderness Area some thirty miles east of Salt Lake City. The house was set against the side of a hill amid aspens and pines, and the front porch gave a broad view across the valley.

All the guests were Mormons, as was Spendlove himself and his wife, Carol. There was Boyd Olsen, the former chief of Dugway Proving Ground's Biological Warfare Test Division; Grant Ash, whose B-24 had been shot down over Germany in World War II, and who later became scientific director of the Deseret Test Center; Fred Houston, who'd been all over the Pacific doing large-scale chemical and biological warfare trials; Lowell Griffiths, a microbiologist; Zen Cox; Paul Adams; Dale Parker; and some others.

The midday potluck was taken on two picnic tables, a fabulous lunch of teriyaki chicken, wild rice, salads, sliced tomatoes, relish trays with pickles and olives, followed by deep-dish pumpkin pie, cookies, brownies, and ice cream. Then, inside, because it was getting cool, with even a hint of snow in the air despite its being not quite fall, the guests assembled in the two-story-high front room of the A-frame, for the annual show-and-tell, when each of those present recounted a high point of their lives since the last reunion a year earlier. Paul Adams and his wife, Ruth, had had their roof tar-papered, and one day while Paul was walking around up there he stepped on the tar paper that covered a skylight, fell right through, and nevertheless miraculously emerged unscathed. Alice and Mel Ludlow, who had owned a trout farm in retirement, recently sold the place and moved into a new home. Cliff Spendlove and his wife had just returned from two weeks in China. And so on. Mention was made of those who had departed the scene since the last reunion, including Dorland Alred, who both spoke and sang at his own funeral (on videotape), and then was buried in the coffin he had designed and made with his own hands.

Then, finally, in late afternoon, the former core nucleus of the American biological warfare program's large-scale test series, the masterminds of SHADY GROVE, NIGHT TRAIN, MAGIC SWORD, EAGER BELLE, AUTUMN GOLD, and all the rest, as kindly and gentle a group as you could hope to find anywhere, made their several promises to return again next year, said their good-byes, and drove down the dirt road and out of the canyon.

Two weeks later, over the weekend of September 25–27, 1998, the Camp Detrick Whitecoat volunteers held a reunion in the newly built, grand, and mammoth Seventh-day Adventist Church in Frederick, Maryland. There were some 215 former Whitecoats in attendance, approximately ten percent of the more than 2,000 Army Seventh-day Adventists who had volunteered themselves up for the program during the nineteen years of its existence between 1954 and 1973.

The program started with Vespers on Friday night, with Invocation and singing of the hymn "Marching to Zion," and then a roll call of what Dr. Frank Damazo, the master of ceremonies, referred to as "our super-special guests," the two hundred assembled Whitecoats, who now lined up and came toward the front of the sanctuary, and proceeded to identify themselves by name, rank, years of service, and current occupation. The roll call lasted some three hours, well into the night.

The next day, Saturday, was the highlight of the reunion—not because it was the Sabbath, which it was, but because it was the day that the Whitecoats got to see Fort Detrick again. They boarded a bunch of olive-drab U.S. Army buses just like in the old days and were driven the four or five miles to the post. It was now an open base; all you needed to get in was a driver's license. They toured USAMRIID headquarters, saw the maximum-containment labs, and walked through the "slammer," the facility's two quarantine rooms. Among those taking the tour were four from the 1955 Q fever trials at Dugway: Bill Twombly, Lloyd Long, Wendell Cole, and Louis Canosa.

Also making the rounds at Detrick was Leonard Barnard, who'd earned an extra measure of Whitecoat fame by having been exposed to the Q fever microbe not once but twice, at the 8-Ball, an object he hadn't seen in more than forty years. He wanted to see it again, and so he was driven over to the site. A couple of new buildings partially blocked the view, so you came upon the 8-Ball suddenly, this great silver sphere that loomed overhead like an alien spacecraft. When he finally saw it, all he said was: "Holy smoke."

In 1945, at the end of the war, Mrs. R. G. C. Maitland, the former owner of Gruinard Island, asked the British government for her island back. But since the land remained contaminated (although with what they wouldn't say, it still being an official secret), the Ministry of Supply was forced to refuse. They assured her, however, that if and when the island was ever cleaned up, she could repurchase it for the same amount they'd paid her for it, £500.

The Porton scientists had already failed in their one attempt to rid the place of anthrax spores when, in the late summer of 1943, they set fire to the island. Such a drastic measure, they had thought, would be likely to kill even as hardy a microbe as an anthrax spore, and so starting low near the water they ignited the brush parallel to the shoreline. The summer had been dry and the grass burned easily, the flames shot up the hillside, crept across the spine of the island, and started down the other side. The grass was still burning well after nightfall, the pockets of red smoldering brush visible from the mainland.

But when the scientists later returned to the island and took soil samples, they were amazed to find the anthrax organism still there in abundance, good as new. Apparently the place would remain lethal for the indefinite future, and so the British government formally closed off the island and stationed signs at 400-yard intervals around the island's perimeter that said:

GRUINARD ISLAND

THIS ISLAND IS GOVERNMENT PROPERTY UNDER EXPERIMENT

THE GROUND IS CONTAMINATED WITH ANTHRAX AND DANGEROUS

LANDING IS PROHIBITED

In 1986, forty-three years after the last of the anthrax bomb trials, the scientists finally came up with a disinfection scheme that worked. The plan was to douse the island's hot areas—which proved to be only 3 acres out of the total 550—with a mixture of 5 percent formaldehyde, a sporicidal chemical, in seawater. In small-scale experiments on the island in 1982, such a solution had killed all the embedded anthrax spores.

So in the summer of 1986, a team of British spacemen came back yet once more to Gruinard Island, and with large hoses and irrigation lines they applied the decontaminating solution to the ten acres of land that had surrounded the original bomb site. Then they went away and came back again a year later and repeated the process, after which all tests for anthrax were entirely negative.

The scientists then brought a flock of healthy sheep to the island, let them graze on the grass, and found that they suffered no ill effects. On April 24, 1990, the British Under Secretary of State for Defence came to Gruinard and formally declared the island "fit for habitation by man and beast." A week later, the Ministry of Defence transferred ownership of the island to Mrs. Maitland's heirs, and in 1994 the heirs sold the island to the present owner of Gruinard Estate, a private landowner who lives near Edinburgh.

Except for the birds and wild rabbits and the seals who sun themselves on the shingle peninsula, the island itself remains uninhabited. The shepherd's hut and the stone animal pen that held the sheep prior to the original anthrax tests are much the same as they were fifty years ago. The tall gray cairn at the hilltop still looks out over Gruinard Bay. And on the west side of the island, just above the waterline, the crushed bones of the dead sheep still lie buried under the Gruinard rock.

Acknowledgments

his book would not have been possible in its present form without the benefit of the Freedom of Information Act (FOIA) in the United States, and the Access to Information Act (AIA) of Canada, whose services provided me with more than 2,000 pages of formerly confidential, secret, and top secret government documents. For their help in locating and supplying copies of these materials, I am indebted to Teresa S. Shinton (U.S. Army Dugway Proving Ground); Cheryl S. Fields (U.S. Army CBDCOM, Aberdeen Proving Ground); Lt. Col. Pauline Cilladi-Rehrer (USAMRMC, Fort Detrick); Lee S. Strickland (Central Intelligence Agency); and Dr. John Clearwater (National Defence Headquarters, Ottawa).

For their help in supplying additional source materials, I am grateful to Jane August (History Office, Wright-Patterson Air Force Base); Lt. Heather Shively (USAMRMC, Fort Detrick); Gail Bishop and Susan Allen (Gulf Islands National Seashore, Mississippi District); Thomas R. Freeman (U.S. Army Corps of Engineers, St. Louis); and Carrie Coberly (Office of U.S. Senator Edward M. Kennedy). Particular thanks to Richard Pastorett (Fisher Library, U.S. Army Chemical School, Ft. McClellan), who provided me with a clean copy of an especially important historical document.

For their personal assistance in researching this book I am indebted to John Bellshaw at Inverness and Dundonnell, Scotland; Brian Eady at Badluchrach jetty and Gruinard Island; and J. Clifton Spendlove at Salt Lake City, Fort Douglas, and Weber Canyon, Utah.

For their hospitality and kindness at Headquarters, U.S. Army Garrison, Fort Detrick, I am heavily obligated to Norman M. Covert (Command Historian) and to Janice Krauss. Also in Frederick, Maryland, I would like to thank for their help Eric Olson, Frank S. Damazo, M.D., Peter B. Jahrling, Robert Peel, Hubert Kaempf, John Bennett, and Riley D. Housewright.

For assistance that went far beyond the call of duty I would like to extend appropriate thanks to Jeffrey K. Smart (Command Historian) and James J. Valdes (Chief Scientist for Biological Sciences, and Scientific Advisor for Biotechnology, U.S. Army CBDCOM, Aberdeen Proving Ground); Gradon B. Carter at CBDE Porton Down, England; and to John Dudley Morton.

For color slides and prints of Howland Island, Baker Island, and Johnston Atoll in the Pacific, thanks to Rick Steiner (University of Alaska) and Mark Rauzon (Berkeley, California); and for prints of the thirty original Whitecoat volunteers at Andrews Air Force Base, many thanks to George DeMuth, M.D.

For his exceptional detective work in finding a small news item concerning the suicide of Arvo T. Thompson, my hat's off to Masahiro Yamada (Library, *Pacific Stars and Stripes*, Tokyo).

For copies of documents concerning the Korean War allegations I am grateful to Conrad Crane (United States Military Academy, West Point) and to Milton Leitenberg (University of Maryland).

Four authors of previous books about biological warfare kindly provided me with documents, contacts, and advice, for all of which I would like to thank Wendy Barnaby, Leonard A. Cole, Sheldon H. Harris, and Seymour M. Hersh.

This book owes a substantial share of its existence to two Bill Patricks: William Patrick, former senior executive editor at Henry Holt, who provided support and criticism throughout; and William C. Patrick III, former chief, Product Development Division, Fort Detrick, who furnished much valuable information during the course of several interviews at his home. My thanks to both, as well as to Holt editor David Sobel for suggestions that improved the text at many points and to Lisa Goldberg for her dazzling copyediting.

Special thanks to Hank P. Albarelli, Ray M. Hawley, Chuck Hibbard, Martin Hugh-Jones, Kenneth D. Jones, Jane Maclay, Jean V. Naggar, Clair B. ("Boyd") Olsen, Graham S. Pearson, Richard Preston, and Pam Regis (to whom I am indebted for the concept of biological warfare as "green" warfare). I would also like to recognize the help of a few individuals who wished to remain anonymous.

Selected Sources

History and Nature of Biological Warfare Research

Published sources

Barnaby, Wendy. *The Plague Makers: The Secret World of Biological Warfare.* London: Vision Paperbacks, 1997.

Bernstein, Barton J. "The Birth of the U.S. Biological-Warfare Program." *Scientific American* 255 (June 1987): 116–121.

Brophy, Leo P., Wyndham D. Miles, and Rexmond C. Cochrane. *The Chemical Warfare Service: From Laboratory to Field.* United States Army in World War II. The Technical Services. Washington, D.C.: Office of the Chief of Military History, Department of the Army, 1959.

Bryden, John. *Deadly Allies: Canada's Secret War 1937–1947.* Toronto: McClelland & Stewart, 1989. (The story of the Canadian germ warfare project.)

Christopher, George W., et al. "Biological Warfare: A Historical Perspective." *Journal of the American Medical Association* 278, no. 5 (August 6, 1997): 412–417.

Cole, Leonard A. *Clouds of Secrecy: The Army's Germ Warfare Tests over Populated Areas.* Totowa, N.J.: Rowman & Littlefield, 1988.

———. *The Eleventh Plague: The Politics of Biological and Chemical Warfare.* New York: W. H. Freeman, 1997.

———. "The Poison Weapons Taboo: Biology, Culture, and Policy." *Politics and the Life Sciences* 17 (September 1998): 119–132.

Covert, Norman M. *Cutting Edge: A History of Fort Detrick, Maryland.* 3rd edition. Fort Detrick, Maryland: Headquarters, U.S. Army Garrison, 1997.

Fox, Leon A. "Bacterial Warfare: The Use of Biologic Agents in Warfare." *Military Surgeon* 72 (March 1933): 189–207; reprinted in Vol. 90 (May 1942): 563–579.

Harris, Robert, and Jeremy Paxman. *A Higher Form of Killing: The Secret Story of Chemical and Biological Warfare.* New York: Hill and Wang, 1982.

Hersh, Seymour M. *Chemical and Biological Warfare: America's Hidden Arsenal.* Indianapolis and New York: Bobbs-Merrill, 1968.

McDermott, Jeanne. *The Killing Winds: The Menace of Biological Warfare.* New York: Arbor House, 1987.

Patrick, William C. "A History of Biological and Toxin Warfare." *Director's Series on Proliferation.* Livermore, Calif.: Lawrence Livermore National Laboratory, May 23, 1994.

Stockholm International Peace Research Institute (SIPRI). *The Problem of Chemical and Biological Warfare.* Stockholm: Almqvist & Wiskell, 1971. Six volumes. (The canonical reference work in English on all aspects of biological and chemical warfare.)

Unpublished U.S. government documents

Cochrane, Rexmond C. *History of the Chemical Warfare Service in World War II (1 July 1940–15 August 1945). Volume II: Biological Warfare Research in the United States.* U.S. Army Chemical Corps: Historical Section, November 1947. (Declassified top secret manuscript; 562 pages, plus appendices, photographs, and annex.)

Department of the Army. *U.S. Army Activity in the U.S. Biological Warfare Programs.* Volumes I and II. 24 February 1977. Unclassified. (Reprinted in U.S. Congress. Senate. Hearings before the Subcommittee on Health and Scientific Research of the Committee on Human Resources. *Biological Testing Involving Human Subjects by the Department of Defense, 1977.* 95th Congr., 1st sess., March 8 and May 23, 1977: pages 22–234.)

Miller, Dorothy L. *History of Air Force Participation in the Biological Warfare Program, 1944–1951.* Wright-Patterson Air Force Base: Historical Division, 1952. (Redacted edition of top secret typescript; 100 pages.)

———. *History of Air Force Participation in the Biological Warfare Program, 1951–1954.* Wright-Patterson Air Force Base: Historical Division, 1957. (Redacted edition of top secret typescript; 267 pages.)

Schertz, Capt. Frank M. *Biological Warfare.* U.S. Army Chemical Warfare Service, 1943. (Declassified secret manuscript; 241 pages, plus indexes.)

War Research Service. *Historical Report of War Research Service, November 1944–Final.* [1945] (Declassified top secret typescript; Archives, National Academy of Sciences.)

Special Topics

Aerobiology

Jemski, Joseph V., and G. Briggs Phillips. "Aerosol Challenge of Animals," in William I. Gay, ed., *Methods of Animal Experimentation.* Volume I. New York: Academic Press, 1965.

Rosebury, Theodor. *Experimental Air-Borne Infection.* Baltimore: Williams & Wilkins, 1947.

Ira Baldwin

Baldwin, Ira L. *My Half Century at the University of Wisconsin.* Madison, Wis.: Omnipress, 1995. (Privately printed oral history.)

"Speech by Dr. Ira L. Baldwin." Fort Detrick Silver Anniversary Luncheon. La Scala Italian Restaurant, New York City. 2 May 1967. (Unpublished typescript.)

Biological munitions
Operational Suitability of a BW Munition. Dugway Proving Ground Report 134 (BW 16-52). Dugway Proving Ground, Utah, 29 January 1954. (Redacted edition of secret document; 90 pages; U.S. Army Dugway Proving Ground Archives. Ft. Belvoir, Virginia: Defense Technical Information Center: AD 366396.)
Tanner, H. G., et al. *Special Report No. 44: Munitions for Biological Warfare.* Two volumes. Camp Detrick, Maryland: Technical Department, Munitions Division, June 1943–Sep 1945. (Unpublished, declassified top secret document. Ft. Belvoir, Virginia: Defense Technical Information Center: AD 310773 [Volume I], AD 310774 [Volume II].)

"Black Maria"
Matousek, James F., et al. *Special Report 42: Engineering, Construction, Operation, Maintenance and Development in Restricted Area.* Section I—General. Section II—"Black Maria." Camp Detrick, Maryland: Technical Department, Engineering Division, 15 November 1945. (Unpublished, declassified top secret document. Ft. Belvoir, Virginia: Defense Technical Information Center: AD 222840 [Section I], AD 222841 [Section II].)

Camp Detrick interrogations of Japanese biological warfare scientists
(1) [Sanders, Murray] *Report on Scientific Intelligence Survey in Japan. September and October 1945. Volume V. Biological Warfare.* General Headquarters, United States Army Forces, Pacific: Scientific and Technical Advisory Section, 1 November 1945. (Declassified secret document; Office of the Command Historian, Headquarters, U.S. Army Garrison, Fort Detrick, Maryland.)
(2) Thompson, Arvo T. *Report on Japanese Biological Warfare (BW) Activities.* Camp Detrick, Maryland: Army Service Forces, 31 May 1946. (Declassified secret document; Office of the Command Historian, Headquarters, U.S. Army Garrison, Fort Detrick, Maryland.)
(3) Fell, Norbert H. *Brief Summary of New Information about Japanese B.W. Activities.* Camp Detrick, Maryland, 20 June 1947. (Declassified confidential document; Office of the Command Historian, Headquarters, U.S. Army Garrison, Fort Detrick, Maryland.)
(4) Hill, Edwin V. *Summary Report on B.W. Investigations.* Camp Detrick, Maryland, 12 December 1947. (Office of the Command Historian, Headquarters, U.S. Army Garrison, Fort Detrick, Maryland.)
[Yoshihashi, Taro]. *Unit 731.* (Recollections by Norbert Fell's translator of his role in the Japanese interrogations, and of Fell's meetings in Tokyo with Russian interrogators. Undated typescript; 16 pages. Office of the Command Historian, Headquarters, U.S. Army Garrison, Fort Detrick, Maryland.)

Camp Detrick pilot plants I and II
Roberts, James L. *Special Report No. 29: Pilot Plant Branch Report on the B.W. Agent "N."* Camp Detrick, Maryland: 1 January 1946. (Declassified secret document; 247 pages; includes photographic plates. National Archives: RG 175, Entry: Biological Department, Fort Detrick, Maryland. Box 2.)

Camp Detrick Special Operations Division

Baldwin, I. L. "Report on Special BW Operations." Washington, D.C.: National Military Establishment, Research and Development Board, 5 October 1948. (Unpublished, declassified top secret document; 11 pages.)

Miscellaneous Publication 7. Study US65SP. Fort Detrick: U.S. Army Biological Laboratories, July 1965. (Unpublished, declassified secret document; 79 pages.)

Miscellaneous Publication 25. A Study of the Vulnerability of Subway Passengers in New York City to Covert Attack with Biological Agents. Frederick, Maryland: Special Operations Division, Commodity Development and Engineering Laboratory, January 1968. (Unclassified.)

Central Intelligence Agency

Hersh, Seymour M. *The Dark Side of Camelot.* Boston: Little, Brown, 1997.

Kalb, Madeleine G. *The Congo Cables: The Cold War in Africa—From Eisenhower to Kennedy.* New York: Macmillan, 1982.

Kronisch v. United States. (Gloria Kronisch, Executrix of the Estate of Stanley Milton Glickman, v. United States of America, Sidney Gottlieb, in his individual and in his official capacities, Richard Helms, in his individual and in his official capacities, and John Does, unknown agents of the Central Intelligence Agency.) United States Court of Appeals for the Second Circuit, August Term, 1997. (http://www.law.pace.edu/lawlib/legal/us-legal/judiciary/second-circuit/test3/97-6116.opn.html)

Report of Inspection of MKULTRA/TSD. July 26, 1963. (Declassified typescript; 42 pages including cover letter and attachments. CIA Archives: MORI DocID: 17748.)

U.S. Congress. Senate. Hearings before the Select Committee to Study Governmental Operations with respect to Intelligence Activities. *Intelligence Activities, Senate Resolution 21.* Volume 1: Unauthorized Storage of Toxic Agents. 94th Congr., 1st sess., September 16, 17, and 18, 1975.

U.S. Congress. Senate. *Alleged Assassination Plots Involving Foreign Leaders.* An Interim Report of the Select Committee to Study Governmental Operations with respect to Intelligence Activities. 94th Congr., 1st sess., November 20, 1975.

"Summary Report on CIA Investigation of MKNAOMI." In U.S. Congress. Senate. Hearings before the Subcommittee on Health and Scientific Research of the Committee on Human Resources. *Biological Testing Involving Human Subjects by the Department of Defense, 1977.* 95th Congr., 1st sess., March 8 and May 23, 1977: pages 244-255. S. Report No. 94-465.

Deseret Test Center

Buhlman, Ernest H. *Test 64-4—SHADY GROVE. Final Report.* Report No. DTC 644115R. Department of the Army. Fort Douglas, Utah: Deseret Test Center, June 1966. (Redacted edition of secret document; 303 pages; U.S. Army Dugway Proving Ground Archives. Ft. Belvoir, Virginia: Defense Technical Information Center: AD 500839.)

Burnes, Robert F., and H. Allen Evans. *Weapon System Effectiveness for Test 68-50.* Los Angeles: Booz-Allen Applied Research, Inc., February 7, 1969. (Redacted edition of secret document; 54 pages; U.S. Army Dugway Proving Ground Archives. Ft. Belvoir, Virginia: Defense Technical Information Center: AD 505262.)

Morrison, John H. *DTC Test 68-50. Test Report.* Volume I. Department of the Army. Fort Douglas, Utah: Deseret Test Center, March 1969. (Redacted edition of secret document; 44 pages; U.S. Army Dugway Proving Ground Archives. Ft. Belvoir, Virginia: Defense Technical Information Center: AD 500676.)

———. *DTC Test 68-50. Final Report.* Volume II. Department of the Army. Fort Douglas, Utah: Deseret Test Center, April 1969. (Redacted edition of secret document; 200 pages; U.S. Army Dugway Proving Ground Archives. Ft. Belvoir, Virginia: Defense Technical Information Center: AD 501487.)

Spendlove, J. Clifton. *The Time of My Life: A Personal History.* Salt Lake City: LDS Archives, 1994. (Unpublished memoir.)

Summary of Major Events and Problems. United States Army Chemical Corps. Fiscal Years 1961–1962. Army Chemical Center, Maryland: U.S. Army Chemical Corps Historical Office, June 1962. (Declassified secret document, 188 pages: pp. 9–16, 98, 135–138.)

Gruinard Island

Carter, Gradon B. Personal communication, 1998.

———. *Porton Down: 75 Years of Chemical and Biological Research.* London: HMSO, 1992.

Manchee, R. J., et al., "*Bacillus anthracis* on Gruinard Island." *Nature* 294 (19 November 1981): 254–255.

———. "Decontamination of *Bacillus anthracis* on Gruinard Island?" *Nature* 303 (19 May 1983): 239–240.

Porton Experiments Committee. *Green Book.* Porton, England, 1943.

Horn Island

Horn Island CWS Quarantine Station (Jackson Project). Archives Search Report. St. Louis: U.S. Army Corps of Engineers, April 1993. (Unclassified.)

Japanese biological warfare program

Harris, Sheldon H. *Factories of Death: Japanese Biological Warfare 1932–45, and the American Cover-Up.* London and New York: Routledge, 1994. (Includes full citations to archival material; corrects some errors in previous accounts.)

Newman, Barclay Moon. *Japan's Secret Weapon.* New York: Current Publishing, 1944.

Williams, Peter, and David Wallace. *Unit 731: Japan's Secret Biological Warfare in World War II.* New York: Free Press, 1989. (The original British edition, published in London as *Unit 731: The Japanese Army's Secret of Secrets*, by Hodder and Stoughton, 1989, includes a chapter on the Korean War withheld from the American edition published by Free Press.)

Korean War allegations

Cowdrey, Albert E. "'Germ Warfare' and Public Health in the Korean Conflict." *Journal of the History of Medicine and Allied Sciences* 39 (April 1984): 153–172.

Endicott, Stephen, and Edward Hagerman. *The United States and Biological Warfare: Secrets from the Early Cold War and Korea.* Bloomington: Indiana University Press, 1999. (Reviewed by Ed Regis, *New York Times Book Review* [June 27, 1999]: 22.)

International Scientific Commission. *Report of the International Scientific Commission for the Investigation of the Facts concerning Bacterial Warfare in Korea and China.* Peking: World Council of Peace, 1952.

Leitenberg, Milton. *The Korean War Biological Warfare Allegations Resolved.* Occasional Paper 36. Stockholm: Center for Pacific Asia Studies at Stockholm University, May 1998.

————. "New Russian Evidence on the Korean Biological Warfare Allegations: Background and Analysis." Woodrow Wilson Center Cold War International History Project *Bulletin 11* (Winter 1998): 185–199.

Moon, John Ellis van Courtland. "Biological Warfare Allegations: The Korean War Case." *Annals of the New York Academy of Sciences* 666 (1992): 53–83.

"Reds Fail to Halt Epidemic in China." *New York Times* (April 3, 1952): 18.

Rolicka, Mary. "New Studies Disputing Allegations of Bacteriological Warfare during the Korean War." *Military Medicine* 160, no. 3 (March 1995): 97–100.

Rosenthal, A. M. "Reds' Photographs on Germ Warfare Exposed as Fakes." *New York Times* (April 3, 1952): 1, 18.

Weathersby, Kathryn. "Deceiving the Deceivers: Moscow, Beijing, Pyongyang, and the Allegations of Bacteriological Weapons Use in Korea." Woodrow Wilson Center Cold War International History Project *Bulletin 11* (Winter 1998): 176–185.

Medical effects of biological pathogens
Sidell, Frederick R., Ernest T. Takafuji, and David L. Franz, eds. *Medical Aspects of Chemical and Biological Warfare.* (Textbook of Military Medicine, Part I: Warfare, Weaponry, and the Casualty.) Falls Church, Va.: Office of the Surgeon General, United States Army, 1997.

Young, George A., Jr., et al. *Special Report No. 28: Laboratory Branch Report on the B.W. Agent "N."* Camp Detrick, Maryland, 1 January 1946. (Declassified secret document; 192 pages. National Archives.)

Death of Frank Olson
"Eight Stories on Frank Olson." Walkersville, Md.: Satellite Video Production, 1997. (Collection of eight television news broadcasts pertaining to the death of Frank Olson.)

Marks, John. *The Search for the "Manchurian Candidate": The CIA and Mind Control.* New York: Times Books, 1979.

U.S. Congress. Senate. Hearings before the Committee on Labor and Human Resources. *Biomedical and Behavioral Research, 1975.* 94th Congr., 1st sess., September 10 and 12; November 7, 1975. (Pages 1005–1132 reprint declassified eyes only, secret, and other CIA documents pertaining to the Olson case.)

Pacific Ocean Biological Survey Program
Gup, Ted. "The Smithsonian Secret." *Washington Post Magazine* (May 12, 1985): 8–20.

Humphrey, Philip S. "An Ecological Survey of the Central Pacific." *Smithsonian Year 1965.* Washington, D.C.: Smithsonian Institution, 1965.

Wickham Steed
Hugh-Jones, Martin. "Wickham Steed and German Biological Warfare Research." *Intelligence and National Security* 6, no. 4 (1992): 379–402.

Steed, Wickham. "Aerial Warfare: Secret German Plans." *The Nineteenth Century and After* 116, no. 689 (July 1934): 1–15; no. 691 (September 1934): 331–339.

Death of Arvo T. Thompson
"MC Officer Found Dead in Tokyo Hotel Room." *Pacific Stars and Stripes* (May 18, 1951): 1.

Whitecoat volunteers
Blodgett, William A., Floyd I. John, and John F. Schindler. *Response of Man in a BW Field Test, Operation "CD-22," BW 3-55.* Dugway Proving Ground Report 186. Dugway Proving Ground, Utah, 17 December 1956. (Declassified secret document; 76 pages; U.S. Army Dugway Proving Ground Archives. Ft. Belvoir, Virginia: Defense Technical Information Center: AD 116869.)
Mole, Robert L., and Dale M. Mole. *For God and Country: Operation Whitecoat: 1954–1973.* Brushton, N.Y.: Teach Services, 1998.

Index

Where the U.S. Army Chemical Warfare Service identified a biological warfare agent, class of agents, or disease by letter code, the code is given in parentheses following the name of the item, e.g., *Pasteurella pestis* (LE).